Circuits I Have Known

Simplified Explanations of Analog Circuits

The Nuts-and-Bolts Approach

With Personal Observations

by

Ronald W. Parker, P.E.

Library of Congress Catalog Number
LCCN: 2009909427

Printed in the United States of America

FIRST PRINTING – 2009

City of Publication
Clinton, CT 06413

SPRINGBOARD

ISBN-10: 0615312373
ISBN-13: 978-0-615-31237-8

Available on Amazon.com

To Julie and Gregory
Thank you for your
complete support

Table of Contents

Observations

In the Beginning…

In the late 1950's and early 60's most of the circuitry was what we today know as 'Analog'. The vacuum tube was king. But it wasn't long before Mr. Shockley's transistor was making breakthroughs. The RCA Tube Manual was the bible of the day. However, a new bible was on the horizon. It was called the G.E. Transistor manual. This was followed shortly thereafter by the logic circuit. ANDs, NANDs, NORs ORs and OREs were the order of the day. IBM mainframes were made of many, many simple logic circuits. When it came to programming, assembly language and 'basic assemblers' were the order of the day. Every month there were new innovations in increased logic density and software compilers. For the logic designers the new bibles were the Texas Instruments 'TTL Handbook' (Volume 1) followed shortly by the 'Motorola CMOS Book'.

But almost at the same time Operational Amplifier came into the spotlight. National Semiconductor seemed to lead the way with a whole series of analog circuits. They had their own bibles known as the 'National Semiconductor Data Books'. Analog Devices company was not far behind. The op-amp opened a whole new world of design. Then, the op-amp got better and better, every month. There was an attempt to merge the analog and digital world. This was done through the FET switch. Using the FET switch, an analog circuit could be gated just like a logic circuit. It could be synchronized with a timing and gating pulse to produce analog signals that were digitally gated.

So what could top this? Well, along came the microprocessor. The first ones were 4-bit and 8 bits jobs, but shortly thereafter came the 16 and 32 bit processors. Manufacturers started incorporating a lot of features into the micros, such as timers, counters, I/O interfaces, A/D and D/A converters, and ultimately the kitchen sink. These became known as microcontrollers. In the mean time, the op-amps were getting better and better in every parameter. The analog/digital WAR was going hot and heavy.

At about this time, the CPLD (Complex Programmable Logic Device) came upon the scene. PALs (Programmable Array Logic) had been around for a while, but they were primarily used to reduce the combinatorial logic, such as NANDs and NORs. Although still a digital logic device, the CPLD lent itself to decision making, much as a microprocessor or microcontroller did, although in a greatly reduced capacity. Its claim to fame was the 'state machine', or decision-making portion in combination with combinatorial or "popcorn" logic that seemed always to be required in most digital application whether microcontroller-based or not.

Many followers of the 'mixed logic' application embraced this new 'decision making and timing generating' device. However, the result was the formation of the basis for the two great electronic religions, the ANALOGIANS (ana-lo-g-en) and the DIGITALIANS. Only because there are certain things that each discipline can do that the other can't is the battle still taking place today, although greatly reduced in intensity. Analogians have a modest respect for Digitalians and Digitalians have a slightly higher regard for Analogians. This is only because there has always been somewhat of an undeserved mystique attached to analog, probably because it has an infinite number of states.

I have neglected to discuss the creation of what has turned out to be the greater part of digital engineering, namely programming. My reason for neglecting this group is that most programmers are not concerned with the nuts and bolts of electronic hardware but rather float above all of this fray in an ethereal mindset that divorces the engineer from reality. Case in point, C++. The few programmers that write application software for a piece of hardware experience the joy of actually making something work, but they're becoming few and far between.

During my career in electronics I have gained expertise in both analog and digital hardware and a basic knowledge of programming. I have also had my share of difficulty in understanding some of the aspects of electronic circuit design. Because of this, I have had to take a 'nuts-and-bolts' approach to understanding circuits. The following circuit explanations are meant to cover many fundamental circuits that I have found useful. The explanations are, as mentioned, nuts-and-bolts. Only when absolutely necessary are any mathematical explanations, other than basic algebra, trigonometry and minimal calculus used in the circuit descriptions.

I hope that you enjoy some of these explanations and get a better 'feel' for electronics rather than what's left after the usual stiff and stilted presentations.

The Author.

An Operation Amplifier Circuit with Gain and Offset

The use of an offset current in an op-amp circuit is really quite simple. The whole idea is that a constant current into or out-of the summing node at the input of the amp will produce a constant offset in the output voltage. An operation amplifier circuit that inverts the input and multiplies it by two is shown below. With 5 Volts IN, the voltage OUT is -10 Volts. The summing node is at the '-' input to the amp. Remember that the inputs always want to be at the same potential. Since one is tied to ground, the other will also try to be at ground potential. The only way to keep this point at ground potential is to change the output current. This current always passes through R2, so R2 produces the voltage drop that appears at the output of the amp.

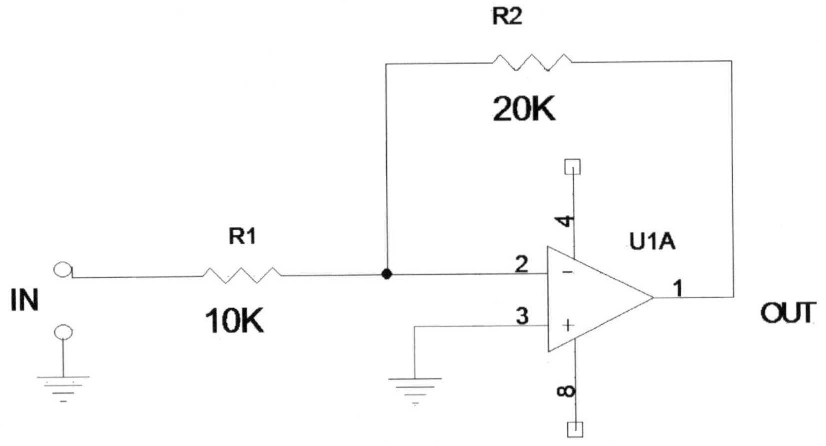

Now lets add an offset current to the '-' input of the amplifier. The resulting circuit looks like this.

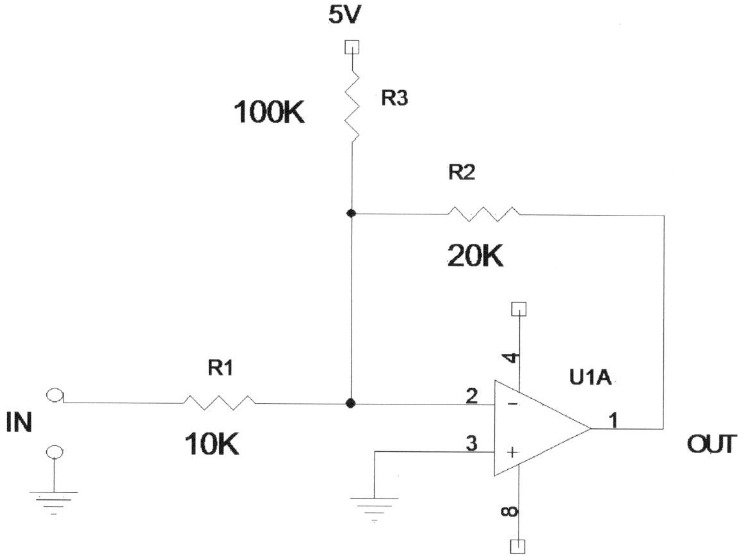

The current produced by the input voltage, IN will be the voltage at IN/10KOhms. In this example let IN = 2 Volts. Therefore the current through R1 will flow from IN toward pin 2 of U1A. This current will be 2/10,000 or 2exp-4 Amps. This same current flows in R2 and causes a voltage drop of 2exp-4 * 20KOhms, or -4 Volts. With one end of R2 at ground, the amp output (OUT) will be at -4 VDC. At the same time, a current flows from 5V, through R3 toward pin 2 of U1A. This current is 5/100KOhms or 50 microAmps. Because pin 2 is always at ground potential, this is a constant current of 50 microAmps. Pin 2 of the op amp is a high impedance input so no current flows into or out of pin 2. The 50 microAmps has to flow through R2 (because U1A pin 1 has the most negative potential). This produces a voltage drop across R2 of 50exp-6 * 20KOhms or -1 Volt (this current flows from Ground to OUT. Therefore it is negative at U1A, pin 1 or OUT). When this voltage drop is added to the -4 Volts the result is -5 Volts OUT.

With a negative voltage IN, the current will be from the node at pin 2 of U1A to IN. Use -2 Volts IN as an example. So the current flowing through R1 will be -2/10KOhms or -2exp-4 Amps. Again, this current flows through R2, only in the opposite direction, keeping the voltage at pin 2 of U1A at Ground potential. This voltage drop is -2exp-4 * 20KOhms or +4 Volts at OUT. At the same time, a current of 5/100KOhms or 50 microAmps flows toward pin 2 of U1Am through R3. This current must be accounted for by the output of the op-amp. It is canceled by a current flowing from OUT to pin 2 of U1A. This current produces a voltage drop across R2 of 50 microAmps * 20KOhms, or -1 Volt. The +4 Volts already across R2 is reduced to +3 Volts.

So the result of the offset current is a constant offset of -1 Volts at pin 1, the OUT pin of op-amp U1A. In any event, the offset is constant. In this example, -1 volt is always added to the output of the op-amp.

How about changing the gain? Let's make the feedback resistor (R2) equal 30KOhms. Ignoring the current contribution from the offset path (R3) for a moment, let the voltage input to the op-amp be +1 Volt. This puts 1 volt across R1 which is still a 10KOhm resistor. So the current into the pin 2 node of U1A is 1/10K or 0.1 milliAmps. Again, this same current flows through R2 which is now 30KOhms so the voltage drip across R2 is -3 Volts (current flows into pin 1 of op-amp U1A). Now let's add in the offset current. This current is still 5V/100,000Ohms, or 50 micro-Amps. The 50 micro-Amps has to flow through R3 also, and produces a voltage drop of 1.5 Volts. This is added to the 3 Volt drop across R2, so the total voltage drop across R2 is 4.5 Volts.

Gain definitely affects the offset. To keep the effect of the offset the same, the best way is to change the value of R1 rather than R2.

Best of luck.

Caught in a Layoff

This is a nice way of putting it. Actually you have been 'selected' out of the corporation. The decision was probably made a long time ago, and by people you knew personally. Sometimes there are many people who know that you're being laid off before you know it. That really hurts. It's like walking around with your head cut off and nobody says 'don't bend over'. Actually, there is nothing that most of them can do. They work for a living too. They are captives of the working system. All they can do is to feel grateful that it isn't them.

However, they can do one thing. They can call you and tell you that it was good working with you. They might even say that if you needed any help, they would be glad to help. O.K., O.K., they may not really mean it, but at least they can say it. If you have any integrity you won't take them up on it anyway. The thing is for them to DO IT. You may want to consider doing this if you are the one that's left standing.

Actually, if a person has been working at the same place for several years (somewhere between 5 and 45 years) they may not believe that they have actually been 'let go' and will have feelings that the company has made a mistake and will shortly realize this and call them back. This could persist for a couple of weeks, or beyond. This feeling is particularly prevalent in people who have believed that the company actually felt that they were important to the company, to its future and to its success, in other words, a 'good employee'.

Believe me, good employees are usually recognized by their bosses and often protected. But during times of pressure ('We need to cut back severely') there is many a 'good employee' who is caught in the net of a lay-off. The best advice I can give to those who are laid off is 'realize fully that you will NEVER see the inside of that work-place again'. (If you are laid off in the afternoon, sometimes you may come back to the work-place, totally escorted, to pick up any belongings you have left, but THAT'S IT!). If a local news-media person contacts you for a story, this will be your only opportunity to tell how you actually fell. The normal reaction is to protect the company for which you worked. Don't worry, that company is GONE. It's 'dead to you'. So you can say what you feel.

In rare instances you may be asked back to share some piece of knowledge that only you have been privy to. This is your opportunity to bargain. Don't sell yourself short. They have just isolated you from the place where you have probably spent more waking hours than anywhere else during that period. Take this opportunity to feed your family and yourself for as long as you can. If it's long enough, this situation may also give you an opportunity to seek other employment. Take full advantage of this. Let's say you are not offered anything. You receive the usual severance package and you have just been escorted out of the facility. Go immediately to the Unemployment Bureau office that is closest to your home. Give them all the information that they request and follow their instructions. There will be checks, whether or not you get a job, there will be checks. These will be paid through your previous employer's insurance, so feel good about this. Now it's time to get another job. See the other suggestions in this book regarding 'finding a job'. Good luck.

Good luck with your next great adventure.

Oscillators

One of the most important circuits required for decision-making circuitry is the oscillator. If you're using a microprocessor, a micro-controller, a state machine or analog circuitry involving timing, not to mention radio frequency transmission and reception, you'll need an oscillator.

There have been all types of oscillators over the course of electronic circuit development. A few of the names are Crystal, Hartley, Colpitts, Phase Shift, and Cathode Coupled Multi-vibrator. The main distinguishing feature of any oscillator is that the output is fed back to the input with an in-phase signal. Believe me, over the years I've created a lot of oscillators without even wanting to.

Most of the oscillators used today are crystal oscillators. They come in standard packages that include temperature stabilization and buffered outputs, and they wind up in your home computer. However, one of the least complicated and least expensive oscillators uses a resistor and capacitor in conjunction with an inverting Schmitt Trigger device. The great thing about this type of oscillator is that you can easily calculate the frequency and that you can vary the frequency with component value changes.

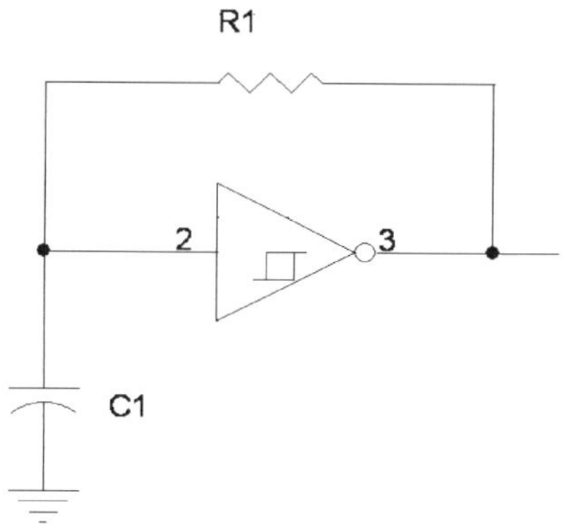

The frequency can vary over temperature, but what are a few cycles between friends if you're requirement is +/- 8% of center frequency. To compute the frequency we have to establish some rules. First, the voltage range of the output from the Schmitt trigger inverter will go from zero to 5 Volts, the same as the Vcc supply to the device. Second, the switching points of the Schmitt input are at 1/3 Vcc and 2/3 Vcc (any input below 1/3Vcc produces 0 Volts at the output and any input above 2/3Vcc produces 5 Volts at the output). So we need to know the charge and discharge times going from 1/3Vcc and 2/3 Vcc. Let's just consider the discharge times first. Using the formula for RC discharge we get:

$$\frac{3.333}{5} = e\text{^}(-t/RC)$$

or $$0.6667 = e\text{^}(-t/(RC)$$

Lets let R = 10KOhms and C = 0.01uF

So $$0.6667 = e\text{^}(-t/(10,000 * 0.00000001)$$

$$0.6667 = e\text{^}\ (-t/0.0001)$$

Taking the log of both sides:

$$-0.40515 = -t/0.0001)$$

Change signs: $0.40515 = t/0.0001$

Solve for 't': $t = 0.0000405$ or 40.5 micro-seconds

This is the time to discharge to 2/3 Vcc.

Remember that the actual time that interests us is the time between 2/3Vcc and 1/3 Vcc. So now let's figure out how long it takes to discharge to 1/3Vcc.

$$\frac{1.6667}{5} = e\text{^}(-t/RC)$$

$$0.3333 = e\text{^}(-t/10,000 * 0.00000001)$$

Take the log of both sides: $-1.8906 = -t/10,000 * 0.00000001$

Change signs: $1.8906 = t/0.0001$

Solve for 't': $t = 189$ exp -6 or 189 micro-seconds

So the time difference is: 189 exp -6 minus 40.5 exp -6 or 144 micro-seconds

How about the Charge time from 1/3Vcc to 2/3Vcc?

Actually, the charge time from 1/3Vcc to 2/3Vcc will be the same as the discharge time for 2/3Vcc to 1/3 Vcc. You can prove this to yourself.

So what is the frequency of the output? It's 'one divided by twice the charge time' (or twice the discharge time), which is:

$$\frac{1}{2 * 144 \text{ micro-seconds}} = 3.47 \text{ KHz}$$

O.K. So what if you want to be able to stop and start this oscillator. Using a Schmitt trigger AND gate you can accomplish this. The only problem is that you don't know at what point in the cycle you stopped or started. You need a little logic in front of the gating circuit to insure that you're stopping and starting at the point you want.

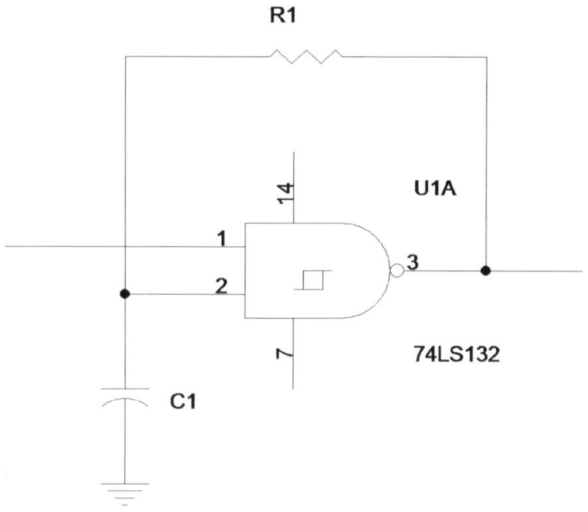

Whoops! Maybe that wasn't such a great idea. Whenever the gating input goes low, C1 will wind up charging all the way up to Vcc and have to charge from that point to 1/3 Vcc. I guess it's better to let the oscillator free run and do the gating downstream.

So much for that! What if you want an a-symmetrical square wave oscillator or, let's say a pulse generator. How about the circuit below:

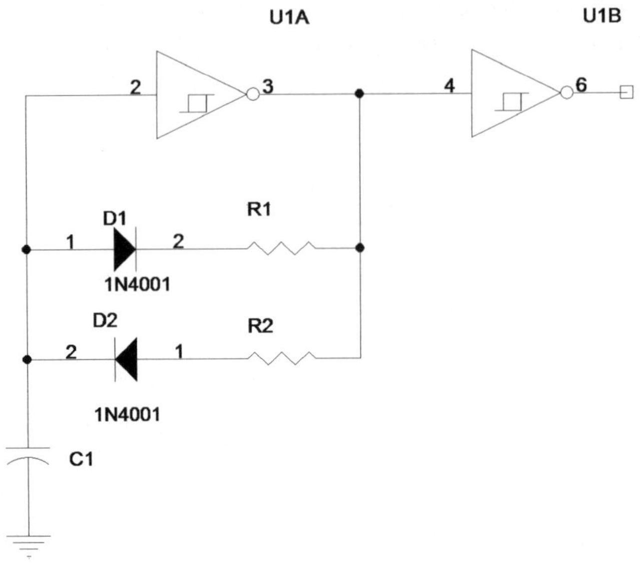

Half of the cycle depends upon R1 and C1 (minus the voltage drop of D1) and the other half of the cycle depends upon R2 and C1 (minus the voltage drop of D2). If you make R1 and R2 variable, you can control both the pulse width and the frequency between pulses. Only problem is, if you change the pulse width you change the frequency and vice versa. Hey! If you want cheap-and-dirty you can't have everything.

Here is one more circuit that is useful when driving a crystal.

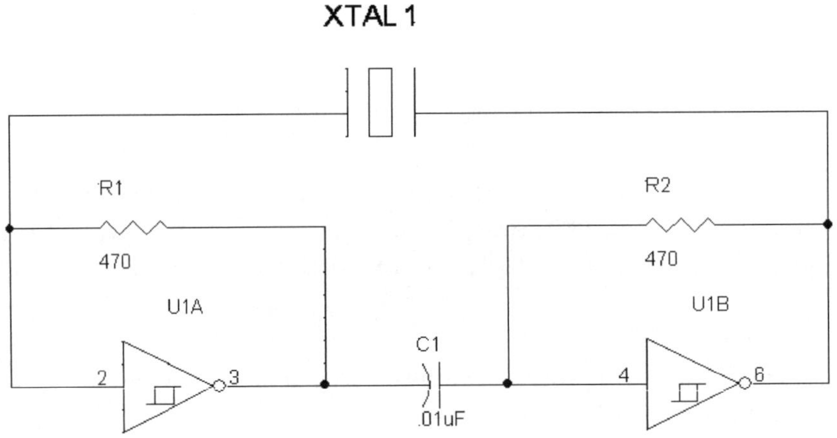

If you have a crystal in the 1MHz to 10MHz range, this should oscillate quite nicely. Be sure to use Schmitt Trigger input devices from the 74 hundred family (running on a 5V supply). No real math involved here. For higher frequencies use lower value resistances. Be sure to buffer the output.

As always, good luck.

Why Analog?

Every time I've run into an engineer, a digital engineer not necessarily a software engineer, there has always been an underlying wish to better understand analog. Maybe it's the school system, maybe it's just that they feel this is what an engineer should know, but almost to-a-man, engineers want to know analog. Well, my story is that I started with analog. Digital came along and I embraced it entirely. But, like all those other engineers, I still wanted to know analog. So I stayed with it through the op amp revolution and evolution. Right now I am fairly comfortable with discrete transistor design right through CPLD coding and micro-computers. I always felt it was good to have as many ways as possible to solve a problem.

So why does every hardware engineer want to know analog better? Because it's the closest thing to reality. Oh, I know that digital sample rates are beyond belief now, but isn't it nice to know that you can solve some simple problem with a few transistors or op-amps and not have to revert to a PIC device or some other controller just to delay the turn-on of a relay or create a linear sweep.

In engineering, it seems that analog is like being 'close to the earth'. Boy, does that sound funny! Natural technology. Nothing like it. Anyway, to me it has always been more fun to solve a problem with analog than digital. That doesn't mean that I don't know when the hand-writing is on the wall and I have to go digital to solve some problem. But I'll always give it a shot with an analog solution before I revert to something that needs programming. In most cases, the solution is a mix of analog and digital circuitry. There is nothing better for timing than digital. Many of my designs include op-amps, transistors, FET switches and digital timing circuits. Hey, I know a good thing when I see it.

Don't get me wrong. I love designing a state machine and coding a CPLD as much as the next guy. I also will use a micro-controller when forced into it. But the smell of a fresh analog circuit just has a certain allure that I can't resist. I'm a sucker for analog.

Peace

Comparator Circuits

Comparators are strange animals but they've been around for eons. Our old buddy 'hysteresis' is normally used in comparator circuits, same as it is in Schmitt trigger circuits. The whole idea of a comparator is to determine whether or not a signal is above some set point or below some lower set point. A certain voltage turns the transistor on and a lower voltage turns it off. There are two circuit configurations that show the two states of a comparator (on and off). The circuit of one comparator configuration is shown below.

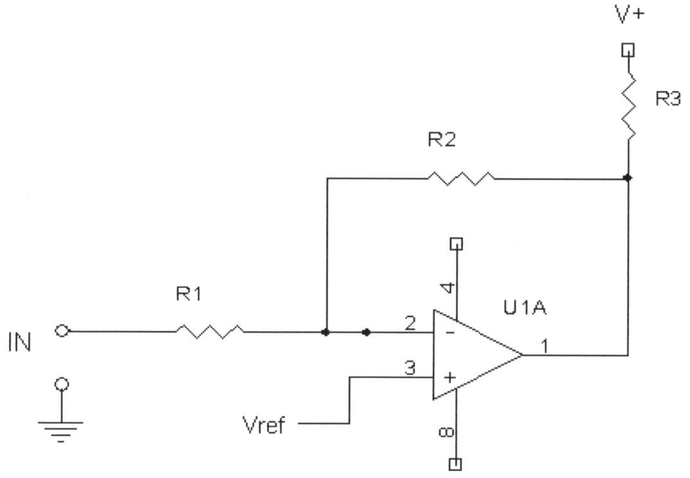

In actuality, the circuit looks more like this one.

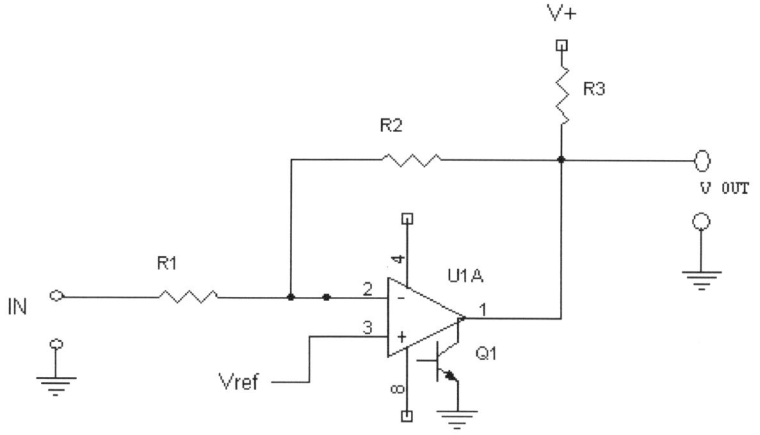

The circuits indicating the switching points are as shown below.

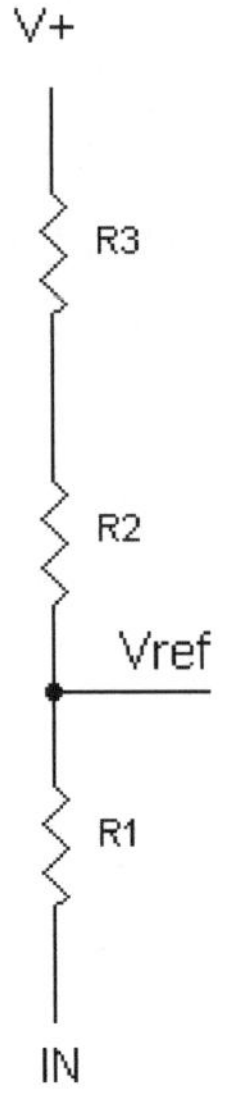

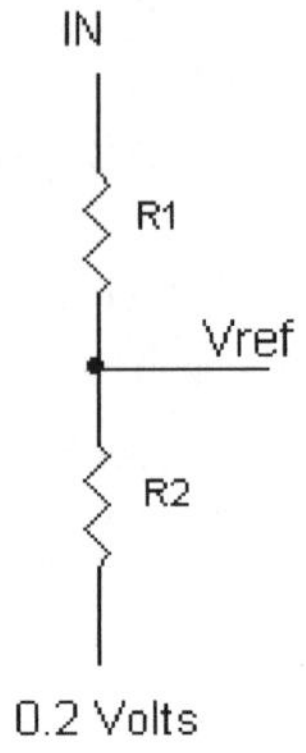

The circuit on the left represents the state in which the output transistor is off. The one on the right indicates the state in which the transistor is in full conduction. The bottom of R2 is at about 0.2 Volts due to the collector-emitter drop of the transistor.

For the sake of explanation, let's give the circuit some values. Actually, we'll use two circuits, each with a different feedback resistor value (R2). The first circuit uses a 1MegOhm value for R2. The second will use 100KOhms. An analysis of differences and similarities will be made during the analysis. The reference voltage for both examples will be 2 VDC.

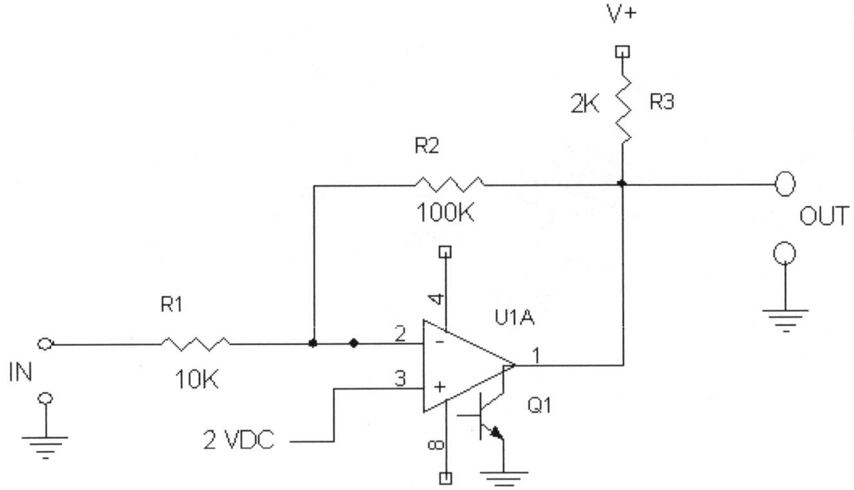

The circuits representing the points at which the output state changes are as follows:

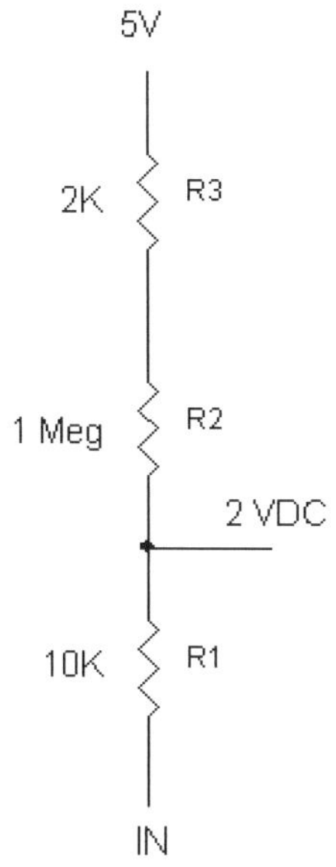

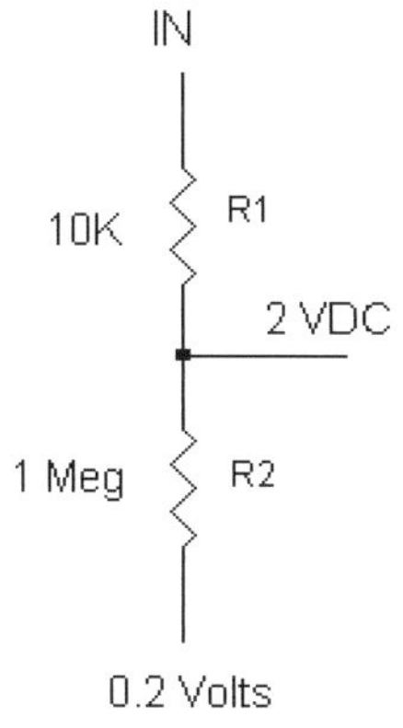

Across the series resistors R2 and R3 there will be 3 volts (5V -2V). The current through this circuit is 3/1,002,000 or 2.994 micro-Amps. Since there is no current drawn at the 2 VDC point (input to the comparator is high impedance), the same current will flow through R1. This makes the voltage drop across R1 to be 2.994 exp -6 * 10,000 or approximately 0.03 Volts. This is subtracted from the 2 VDC reference to give the actual input voltage, which is 1.97VDC
Great! Now we know the voltage at which the comparator will change state when the input is below 1.97 VDC and increasing.

How about the upper limit? Refer to the circuit on the right. The voltage across the 1 MegOhm resistor is 2V – 0.2V or 1.8 VDC. The current through the 1 MegOhm resistor is 1.8/1,000,000 or 1.8 micro-Amps. This same current flows through R1 (10KOhms) and produces a voltage drop of 1.8 micro-Amps * 10,000 or 0.018 Volts. Add this to the 2 Volt reference value and we get 2.018 Volts, the upper switching voltage.

So the switching points are not symmetrical about the reference voltage (approximately 0.02 Volts above the reference and 0.03 Volts below the reference). Keep this in mind when calculating how much margin you need for ‘noise’ and such. Notice that the output low level is about 0.2 VDC because it is the collector output of a transistor that is fully ON. When the transistor is off, the voltage drop across R3 (the 2KOhm resistor) is 2.994 exp-6 * 2,000 or approximately 6 milliVolts. The output of the comparator will be 6 milliVolts below the 5V supply, or 4.994 Volts. Now, how about changing the feedback resistor to 100KOhms.

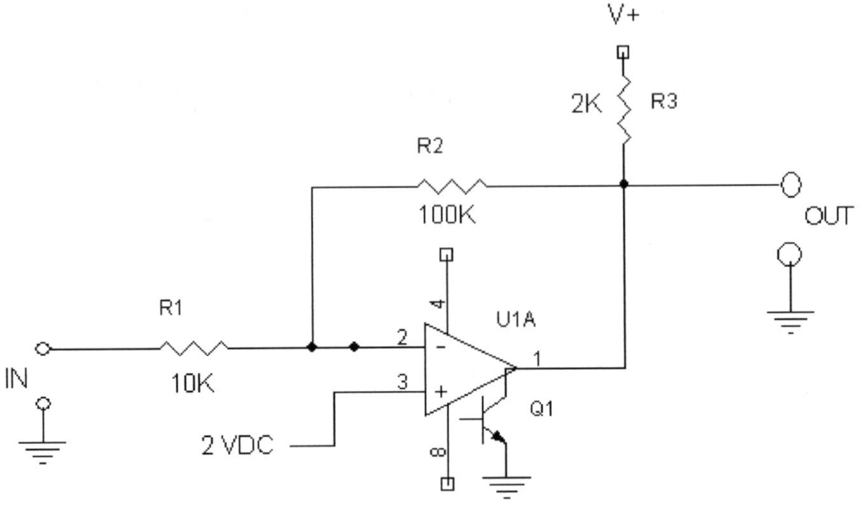

Once again, the circuits below determine the switching points of the comparator.

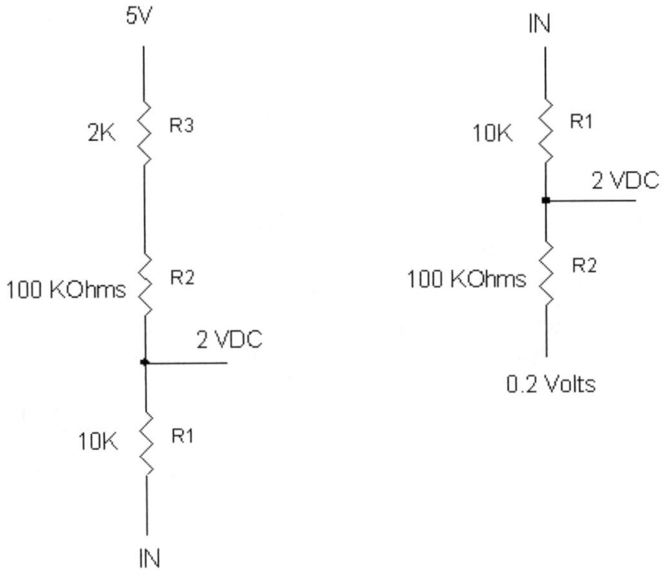

The IN voltage switching points are calculated in the same way as above. In the left circuit, the voltage across R3 and R2 will be 5 – 2 or 3 Volts (once again). The current through R2, R3 and therefore R1 (remember that the 2VDC point is a high-impedance connection to the comparator) is 3/102,000 or 29.41 micro-Amps. So the voltage drop across R1 is 10,000 * 29.41exp-6 or 0.294 Volts. Subtracting this from the 2 VDC reference voltage gives us a lower switching point of approximately 1.7 VDC.

The circuit on the right is shows 2 V – 0.2V or 1.8V across R2. The current through R2 will be 1.8/100,000 or 18 micro-Amps. So, 18 micro-Amps flows through R1. The voltage drop across R1 will be 10,000 * 18exp-6 or 0.18 Volts. This voltage is added to the 2 VDC reference voltage to give a high-level switching value of 2.18 VDC.

So how about the outputs from the two comparator examples? Well, the output voltage of the first circuit (1 Meg feedback) was 6 milliVolts below the 5 Volt supply voltage or 4.99 Volts. In the second case (100KOhm feedback), the voltage drop across the 2KOhm resistor will be 29.41micro_Amps * 2,000 Ohms or approximately 0.06 Volts. The output voltage will be the 5V supply voltage minus 0.06 Volts or 4.94 Volts. Not very much difference, especially if this is going to a high impedance logic circuit input.

Which reminds me, be careful of the load on this circuit as the switching points will be changed by low impedance loads.

Now, suppose you don't have a comparator but you do have a fairly good op-amp available (wide range differential input, high input impedance, high gain, high bandwidth, unity gain stable, low offset voltage,.....) and you want the same characteristics as the comparator you have been using. You might want to look at this circuitry, nasty as it looks.

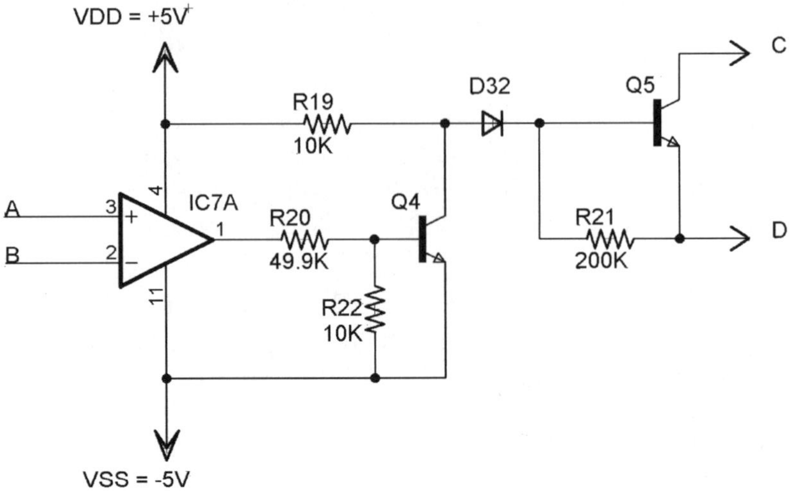

There are both positive and negative supplies for this circuit, although ground is not used. This is made possible by the wide differential input of the op amp. An op amp with rail-to-rail inputs is a possibility. IC7A does the comparing and the output of IC7A swings between VSS and VDD. This output turns transistor Q4 on or off and any current through Q4 passes from VSS, through Q4, through R19 to VDD. R22 and R20 form a voltage divider between the output of IC7A and VSS. If the negative output of IC7A only goes to 2.5V above VSS (in our case, to -2.5V), then the voltage at the junction of R20 and R22 will, with the values given, be about 0.4V above VSS. Q4 needs 0.7 voltage between base and emitter in order to get collector-to-emitter conduction. Since its emitter is at Vss, Q4 will not conduct. So, what about D32? Let's assume that the emitter of Q5 is at Ground Potential. When the output of IC7A is below VSS+0.6V, Q4 will not conduct. The only current path through D32 is from VDD, through R19, through D32, through the base-emitter junction of Q5 to the emitter of Q5 which is at GND potential. So Q5 will turn on, allowing current to flow from 'C', through Q5 collector-to-base, to 'D'. If Q4 is conducting, the anode of D32 will be at the potential of the collector of Q4 which, in our case could be as low as -2.5V. To prevent the base-to-emitter voltage of Q5 from being reverse biased beyond its limits, R21 will conduct any leakage current through D32. The voltage drop across R21 raises the base voltage of Q5 and prevents a high negative base-to-emitter voltage across Q5's emitter-base junction. Note that the voltage on the emitter of Q5 can be below GND potential and can even be as low as VSS. If we want to add hysteresis to this comparator, we do it in much the same way as any comparator. The circuit below shows this similarity.

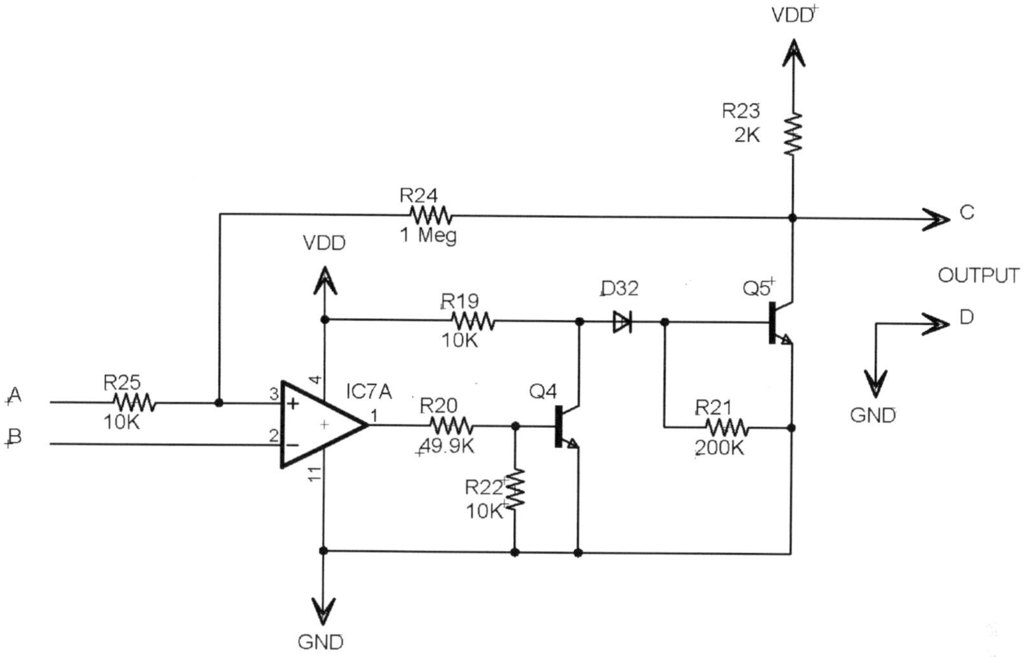

Notice that the 'VSS' supply is now GND and that the switched output is between the collector of Q5 and GND. The hysteresis value is the same as previously discussed for the comparator circuit. If you want to use the VSS supply with the output going between C and D when D is grounded, see the circuit below.

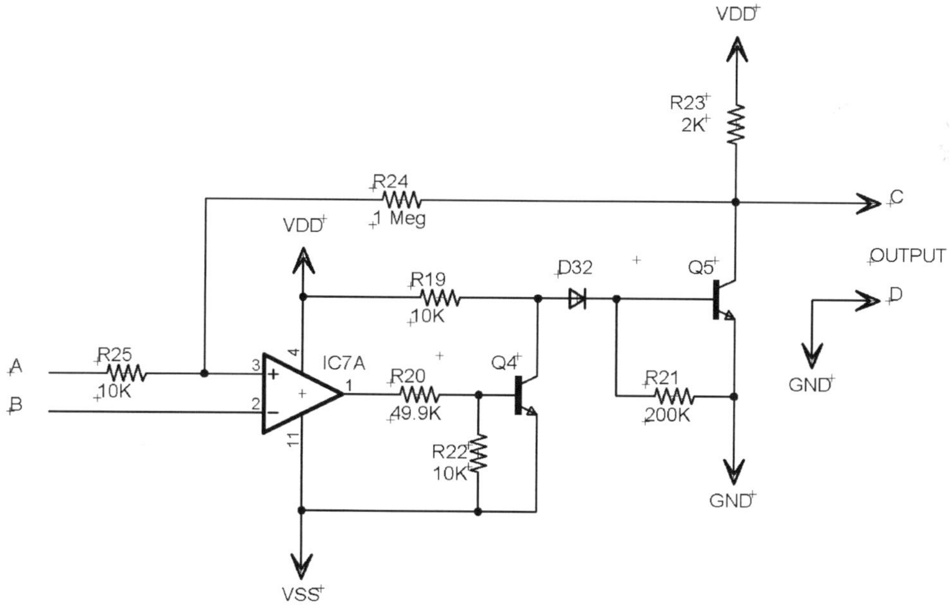

I know, I know. It's a lot of parts. But if you're stuck for a comparator, you can make your own. If you happen to have two op-amps available, the comparator circuit becomes a little simpler. Well, maybe not. But it is another approach.

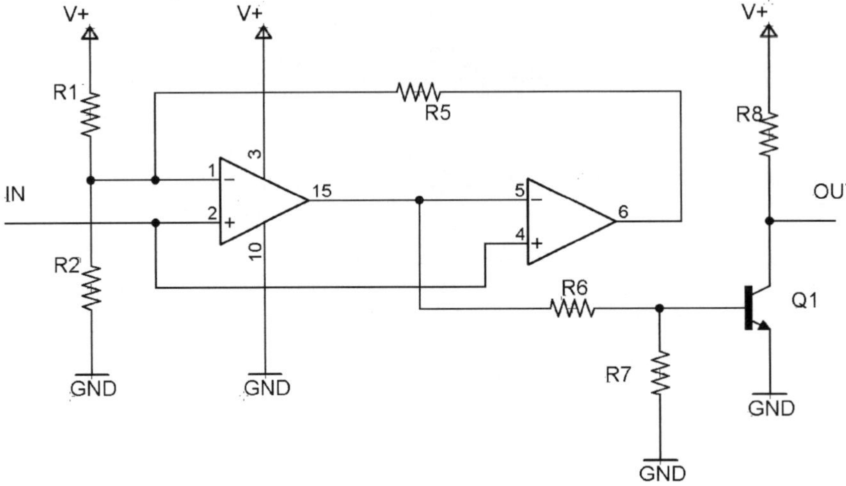

So, what's going on? Resistors R1 and R2 form the voltage divider that determines the switch points for the first op-amp. But, notice that R5 is also in the picture.

Let's make a few assumptions:

- Let R1 = R2 = 100kOhms
- Let R5 = 2MegOhms
- Let the op-amps be true rail-to-rail input and output devices.
- Let V+ = +5V
- Let 'IN' = +1V

Now the voltage on pin 1 becomes V+/2, or 2.5V. Since pin 1 is more positive than pin 2, output pin 15 will be negative. This makes pin 5 more negative than pin 4 and forces output pin 6 to go to +V. The pin 1 input circuit looks like this:

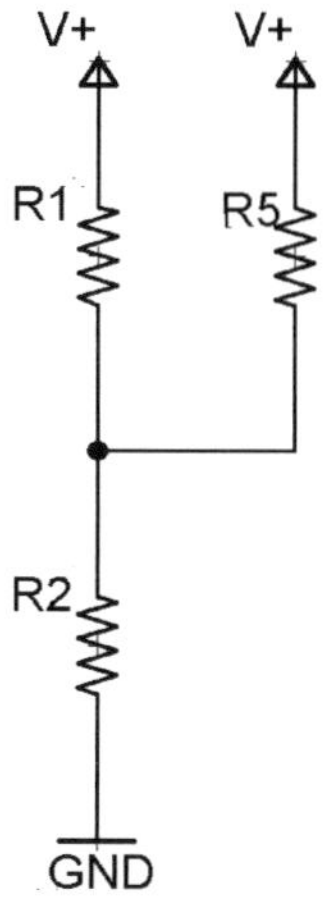

So what's the voltage at the R1/R2 junction (goes to op-amp pin 1)?
First, R1 and R5 are in parallel, so let's represent them as one resistor.

$$\frac{R1 * R5}{R1 + R5} = \frac{10K * 2Meg}{2{,}010{,}000 \text{ Ohms}} = 9{,}950.25 \text{ Ohms}$$

Since this is in series with R2, the total series resistance is:

9,950.25 Ohms + 10,000 Ohms = 19,950.25 Ohms

The current through these resistors is:

$$I = \frac{5V}{19{,}950.25 \text{ Ohms}} = 250.62 \text{ micro-Amps}$$

Multiplying this by the 10kOhm resistance we get:

10,000 Ohms x 250.62uA = 2.506V on pin 1

Now, let's consider that the 'IN' voltage is above 2.506V. This will force output pin 15 to +V and output pin 6 to GND.

The reference circuit becomes:

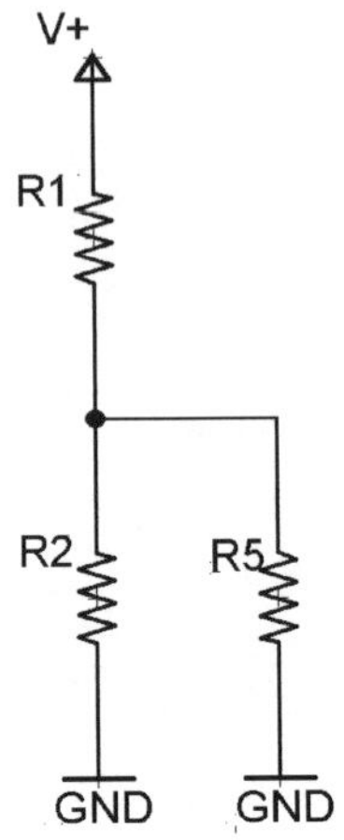

Using the same technique as above, the parallel combination of R2 and R5 has a resistance of 9,950.25 Ohms. The total series resistance is once again 19,950.25 Ohms and the current is 250.62 micro-Amps. The voltage drop of the parallel resistances is:

250.62uA * 9,950.25 Ohms = 2.494 Volts on Pin 1

So when the 'IN' voltage is increasing, it will cause the outputs to switch polarity at 2.506 Volts. When decreasing, it will cause the outputs to switch polarity at 2.494 Volts. With these values, the amount of hysteresis is 0.12 Volts

What about output transistor Q1? It will conduct when pin 15 is positive. R6 and R7 form a voltage divider that keeps Q1 from conducting until pin 15 has transition sufficiently positive to exceed 0.7V at the R6/R7 junction. In this case, R7 really isn't necessary because the voltage at pin 15 is rail-to-rail and, when low will always be below 0.7 Volts. But R6 is definitely necessary in order to keep pin 15 from being clamped to GND through Q1's emitter-base junction.

Best of luck.

Engineer's Egos

So, what do you think is the greatest hindrance to a decent design? It's ego man, it's ego. Engineers, those mild mannered reporters from the Daily Planet, have egos the size of planetoids.

When I started working as an engineer I didn't have all the answers but I was a creative guy. I spent long hours figuring out circuits and working on the best solution to a problem. My problem was that I was the junior engineer. There were other regular engineers, and then there was the chief engineer. I used to get design scraps once in a while, but most of the time what the chief engineer said, or designed, was what wound up in the product.

Most of my satisfaction came from designing my own stuff. I have always designed my own stuff. I've had many successes in terms of having my circuits produced, but the real joy has always been in designing them and making them work. Anyway, eventually I became the chief engineer. The problem with that was that I could relate to the engineers who worked for me. I let them have their heads. Unfortunately, they often made some design errors. Then I had to step in and make an 'observation'. I tried to do this as tactfully as possible, but I inevitably had to offend these engineers. O.K, o.k., I was doing the right thing for the company. It still wasn't comfortable. That darn engineering ego.

The good that I see coming from this is that all engineers take a lot of pride in their work. They're actually artists and like artists, they like to be appreciated and to convey their message through circuitry and programs. I love engineering. I always have. It's on the human cutting edge. I enjoy the people I have worked with, the people I have worked for, and the people who have worked for me. Engineering has been a rewarding and life fulfilling endeavor.

A fellow engineer.

Precision Full Wave Rectifier

It's all about current!
This is a standard, conventional precision full wave rectifier circuit. You may be using this circuit so often that you don't even need to know what's going on, or you've forgotten. Anyway, here it is again. Remember, it's all about current.

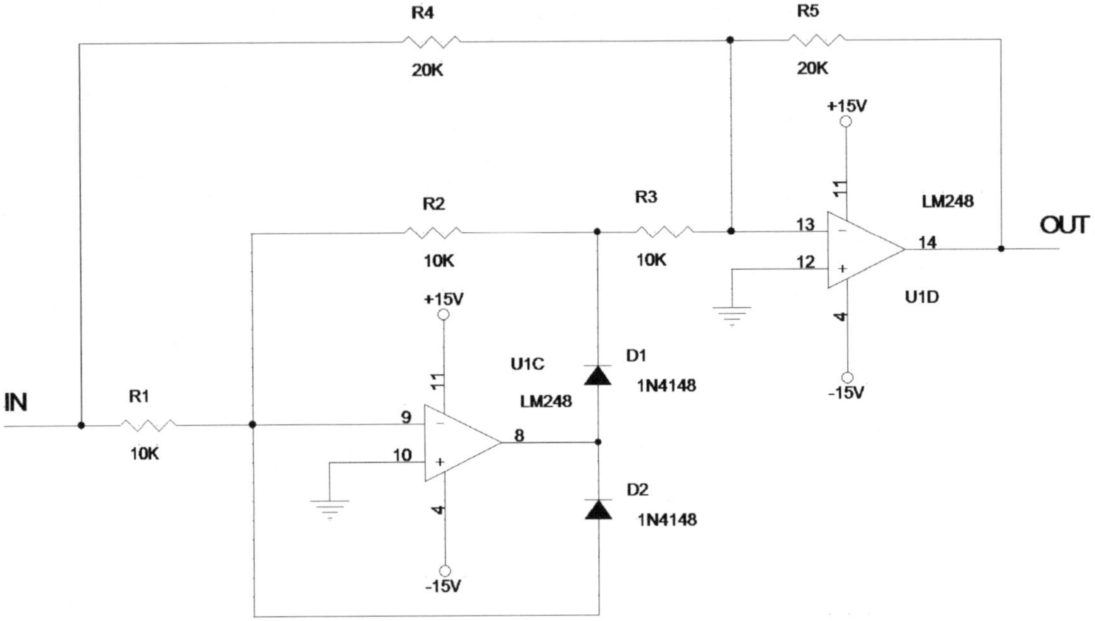

What happens? Well, you put a positive and negative going signal (about ground) on IN and you get OUT only a positive going signal with the negative part of the input signal mirrored above ground. What happens when we put a positive signal in?

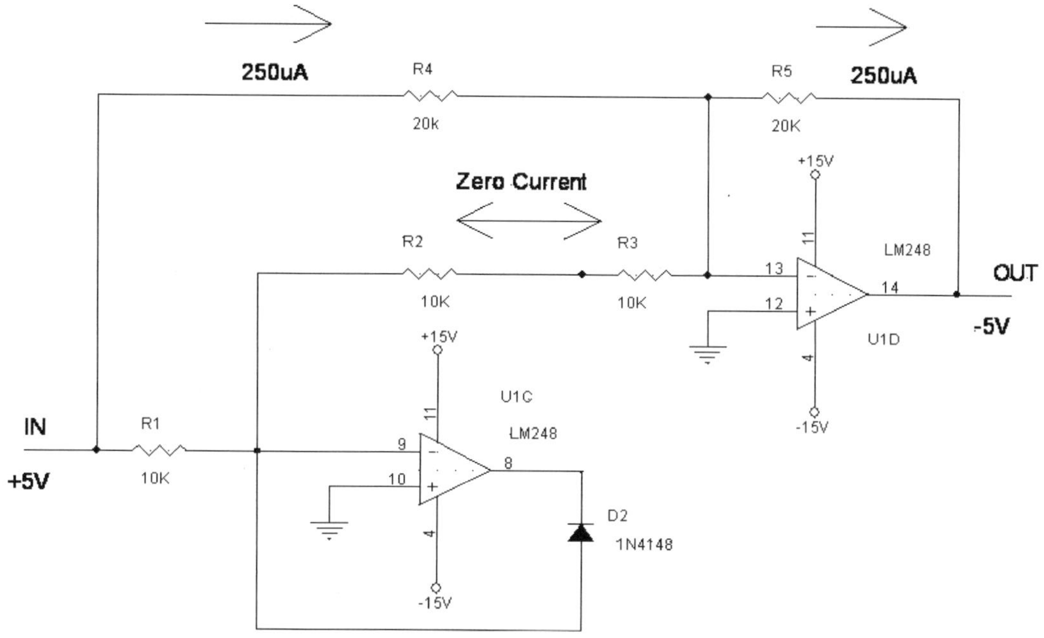

With a positive signal IN, the output of U1C will go negative ('-' input more positive than the '+' input). This effectively takes D1 out of the picture as it is reverse biased (the junction of R2 and R3 have no way of being positive. The '-' input pins of both U1C and U1D want to be at Ground potential). However D2 is fully capable of conducting (forward biased) and pin 9 of U1C goes hard to Ground. With both pin 9 and pin 10 at Ground potential, U1C is happy. So the path from U1C pin 9 to U1D pin 13 conducts no current (no difference of potential) and the circuit reduces to:

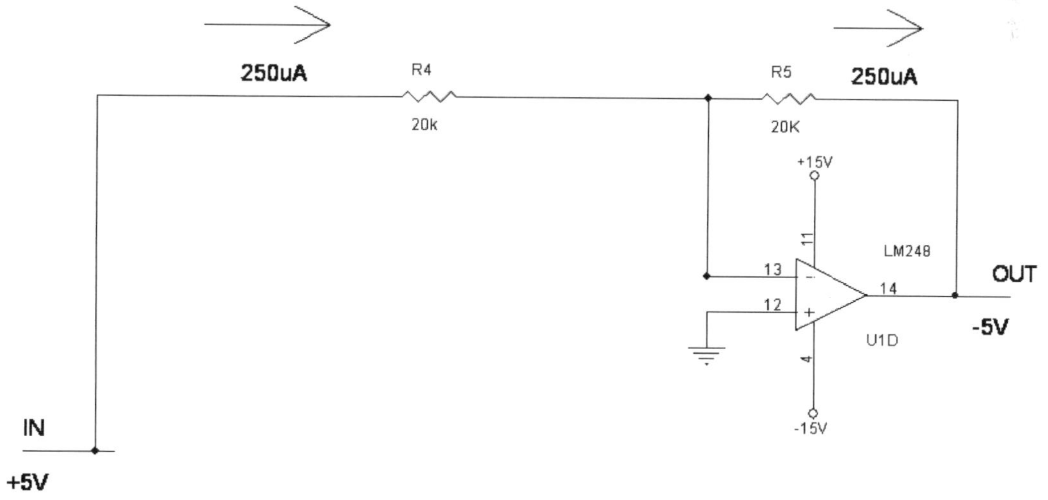

Hey! Wait a minute. I though we said that you put a positive voltage in and get a positive voltage out. OK. OK. I'll explain later. Right now let's see what happens when we put a negative signal IN. From our first figure (above), with a negative signal IN, the output of U1C will go positive. This reverse biases D2 so that D2 has no effect on the circuit and the circuit becomes:

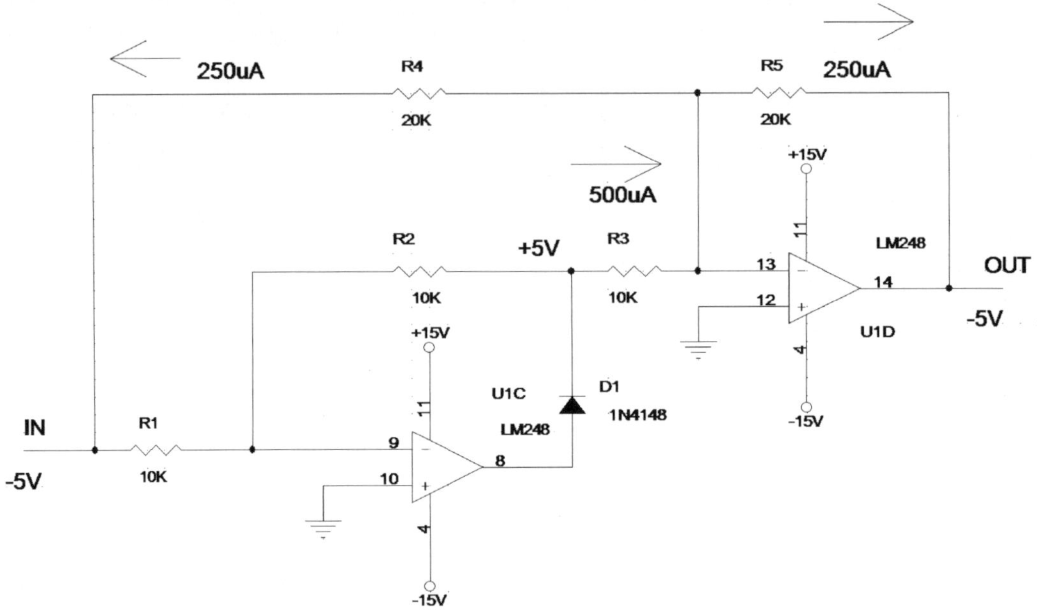

Diode D1 is forward biased and keeps the junction of R2 and R3 at +5Volts. This is because U1C becomes an inverting, unity gain stage. With +5V at one end of R3 and Ground at the other, the current through R3 will be 500uA. We have to account for this current through both R4 and R5. R4 has -5V (IN) on one end and Ground on the other (Pin 13 of U1D), so the current through R4 will be 250uA away from U1D pin 13. Now, to get rid of the remaining 250uA. This still has to be taken away from U1D pin 13. Since U1D is configured as an inverting amplifier and the +5V at one end of R3 is trying to make the inverting input more positive, the output of U1D will go negative. The remaining 250uA will flow through R5 and produce a voltage drop of 5 volts across R5. The current through R5 flows from U1D pin 13 to OUT, so that the voltage at OUT is -5V.

There we go again! The output is still negative. All right, I put the diodes in backwards. If we merely switch polarities on D1 and D2, the output will be positive all the time (check it out). At least we know that we can make a Negative Going Precision Full Wave Rectifier.

Happy circuit designs.

Finding a Job

I want to share something with you about finding a job. I know that engineers are usually engineers because they like to problem solve. Well, here's a problem that may require a few additional skills. Finding a job is actually easier than you think, even in hard times. One of the many times that I was caught in a layoff I wound up pursuing the normal methods of getting a job. I contacted people with whom I had worked before, I job-searched on the Internet, I read the papers and I went to job fairs. Hey, these avenues were exhausted quite rapidly. Pretty soon I was sitting at home waiting for the phone to ring. That is a NO-NO!!!

The whole idea is to approach finding a job as if finding a job is a job in itself. I started by printing up a stack of resumes. I took a pad and pencil so that I could keep track of where I had been. Then I got off my butt, put on a suit and tie and went out knocking on doors. Where? Well, it turned out to be just about anywhere there was an industrial park. I lived in a town that was near an interstate. I think that just about everyone can say that. So I got on the highway, the direction didn't matter, and took the first exit. Lo and Behold there were at least three industrial parks within 1/8th mile of that highway exit. I knocked on every door. I didn't care what their business was, in fact most of the time I didn't even know what business they were involved in. I was dropping off resumes anyplace that would take them. I kept doing this at every exit within a half-hour drive in either direction. This whole scenario went on for about one week. Then I had a call. Someone wanted to interview me. I went to that interview and they tried to seriously "low-ball" me on the money. I didn't accept that job and went back to knocking on doors. Most of the people I left resumes with were receptionists. No problem. Receptionists have their fingers on the pulse of the business they represent. They also deal with sales people, who deal with everyone in the business. The whole idea is exposure. If you hit an industry that you know deals with engineering you may ask to speak with someone in engineering but, in actuality I never needed to do this. After about three and ½ weeks I got a call regarding an engineering position. I did well on the interview, mainly because I knew that I could just keep knocking on doors and I'd get another offer at some point. This gave me a certain degree of confidence in negotiations. Anyway, I took the job. It lasted almost five years, after which the company was sold to the competition.

I had originally estimated that I would have to distribute in the neighborhood of 400 resumes. In actuality it was closer to 200 before I was successful. Actually, when I took the job I felt a little let-down because now I didn't have the job of finding a job anymore and, I felt that I had gotten pretty good at it. So what was the commute? It turned out to be 18 minutes away from my home. Not bad in hard times. I know that not everyone is cut out to approach the door of a business 'cold', but it gets easier and easier with every door and face it, it's better than sitting at home waiting for that darned phone to ring. That could take years.

See you out there.

Operational Amplifier Power Supply

Here's a circuit that buffers and increases the current from a reference source, 5 Volts in this case. At the same time, it's output is double that of the 5V reference source. There is also current limiting included in the circuit.

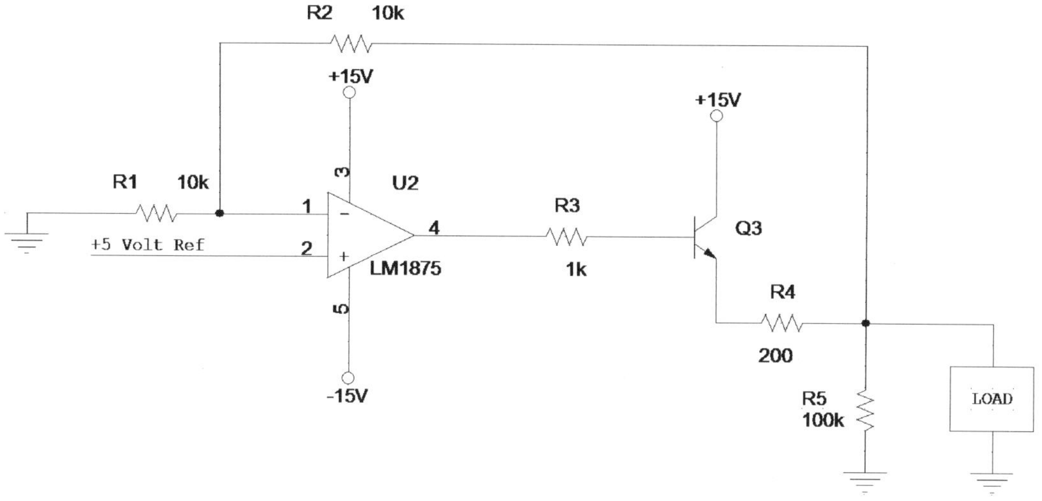

If you look closely at this circuit it is nothing more than an inverting amplifier (two 10KOhm resistors, both at the '-' input node) with a +5V reference voltage on the '+' input. The difference is a bunch of stuff between the output of the op amp and the feedback resistor, R2.

The output of U2 will go high when the '-' input is lower than the '+' input. When U2's output goes high, it turns on Q3. This causes current to flow through R2, R4, R5 and the LOAD. At the junction of R2, R4, R5 and LOAD is the feedback point we were talking about. In order to make the voltage at the '-' input of U2 go higher, the voltage at the R2, R4, R5, LOAD junction has to increase. Actually, it has to increase to +10V in order to satisfy the '-' input of U2. The voltage across R4 is a result of how hard Q3 is turned on. But, we must be careful here. The output of the operational amplifier may not go to the rail. Even the so-called rail-to-rail operational amplifiers, although they may come close to the rail in output voltage, can only source a small amount of current as their outputs approach the +V rail. This small current may not be enough to cause additional conduction in Q3.

Actually, R5 isn't really required. The current path that satisfies U2's inputs is from +15V through Q3, through R4, through R2, through R1 to Ground.

In the event that the voltage at the '-' input of the op-amp goes above the voltage at the '+' input, the output of U2 will go more negative (although still a positive voltage) and force the base of Q3 more negative. So what! The path from the emitter of Q3, through R4, R2 and R1 to ground still exists.

Anyway, we'll assume that U2 can output max current at the +V rail. This means that the emitter of Q3 has a theoretical maximum voltage of 14.3 volts. Remember, we're still trying to maintain 10 Volts at the R2, R4, R5, LOAD junction in order to satisfy U2's inputs. So let's total up the currents when U2 is satisfied. One side of R2 must be at 5V to satisfy U2. This puts 5 Volts across R2 and the current through it becomes 5/10,000 or 0.5mA. This current will always exist in order to keep U2 satisfied. This current has to flow through R4, the 200 Ohm resistor and contributes a voltage drop of 0.1 Volts across R4.

The total voltage across R4 in this example can reach 14.3 -10V or 4.3 Volts. 0.1 Volts is used for feedback, through R2 to the '-' input of U2, so 4.2 Volts is the maximum across R4. This gives us a current of 4.2/200 or 21 milliAmps. O.K. Let's say that we want to limit this current to 10 milliAmps. How are we going to do this? Well, we know that to get 10 milliAmps through R4 the voltage across R4 must be 0.01Amps * 200 Ohms, or 2 Volts. Don't forget the 0.1 Volts through R2, needed to satisfy the input of U2. So the voltage across R4 must be 2.1 Volts.

One thing we could try is to clamp the Q1 emitter side of R4 at 12.1V with a 12.1V Zener diode. BUT LOOK WHAT HAPPENS!

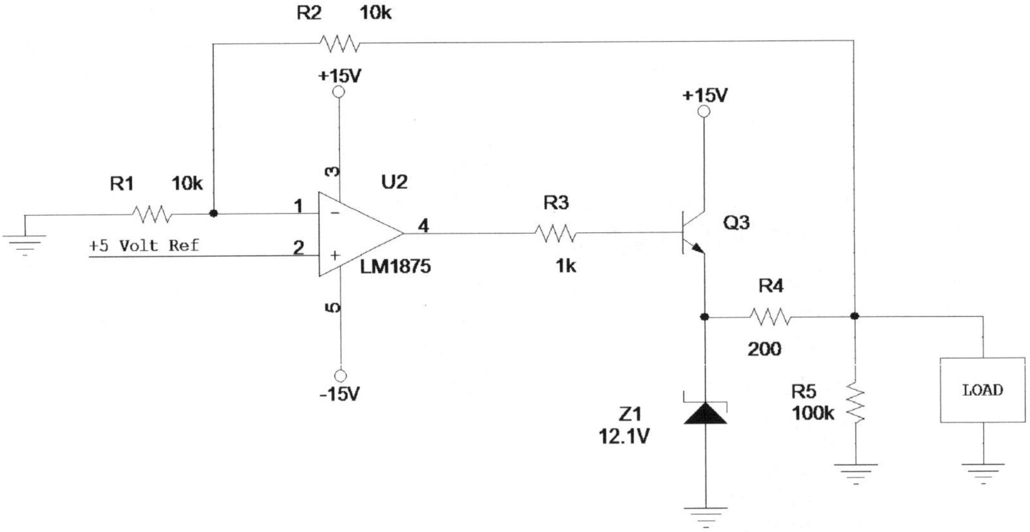

When the load is of a low enough resistance to draw more than 10 milliAmps, the voltage drop across R4 will exceed 2.1 Volts. The 12.1V Zener will begin to conduct. The 12.1V Zener will basically become a short to 12.1 Volts for any voltage above 2.1V across R4. It doesn't take long for the current through the Zener to become significant and eventually to burn up Z1 or Q3, or both.

O.K. So this wasn't such a good idea. How about a resistance in series with the collector of Q1? To keep 12.1 Volts at the side of R4 connected to Q1's emitter, let's consider that Q1 is saturated. That means that the voltage from the emitter to collector of Q1 will be nearly 0.2 Volts. So what's left? What's left is:

15V - (12.1V + 0.2V) or 2.7 Volts. In order to drop 2.7 Volts at 10 milliAmps the resistance at the collector of Q1 must be 2.7/0.010, or 270 Ohms.

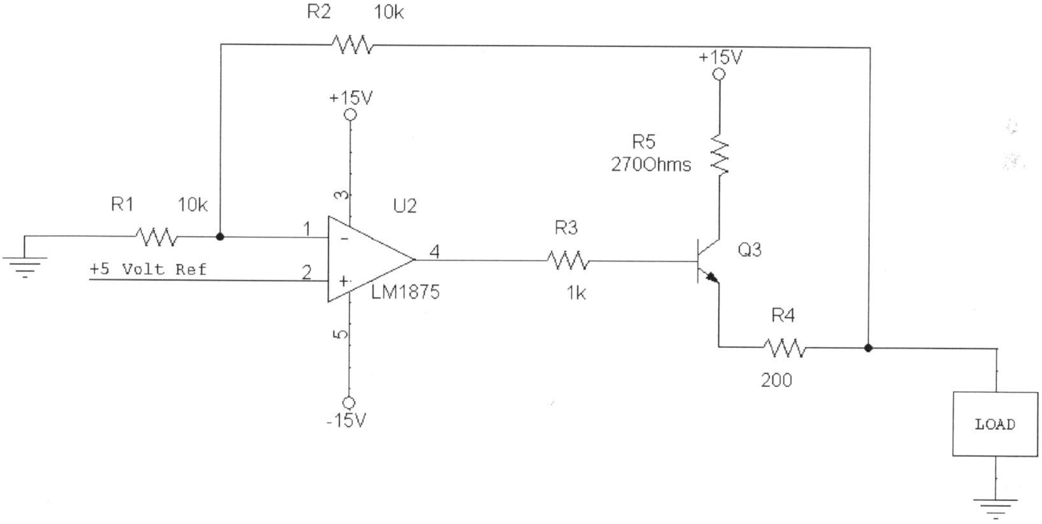

So what will the output voltage of U2 be when the load is shorted? Let's say that the Beta of Q3 is 100. The total current, with Q3 saturated is:

$$+15V / (R5 + R4) = 32 \text{ milliAmps (approximately)}$$

32 milliAmps flows through R4. The voltage drop across R4 is:

$$0.032A * 200 \text{ Ohms} = 6.4 \text{ Volts}$$

A good approximation of Q3's base current is:

Current through Q3 / 100 = 0.032 /100
= 0.00032A, or 320 microAmps.

The drop across R3 would then be:

320uA * 1000 Ohms = 0.32 Volts

Add this to the voltage drop across R4 and to the 0.7V emitter-to-base drop of Q3 and you get:

6.4V + 0.7V + 0.32V = approximately 7.5 V

How nice! BUT THAT'S NOT WHAT'S HAPPENING'!

The output of U2 isn't happily sitting at 7.5 Volts at all. U2's inputs aren't satisfied, so U2 keeps pumping out voltage to try to compensate. It pumps it out until it reaches the limit, which we have stated to be the power supply value, or +15V.

With Q3 totally saturated, there is no place for Q3's emitter-to-collector current to go when there is an increase in the voltage driving the base. Current begins to travel through the emitter-base diode and through R3 to +15V. Here we have to be careful that we don't exceed the base current limit of Q3, although U2 probably isn't capable of sourcing enough current, through R3, to damage Q3.

So, R3 will have a current contribution to the load. With +15V at the top of R3, and ignoring the emitter-base diode drop for the time being, we can consider R3 and R5 to be in parallel.

The parallel combination of R3 and R5 results in a resistance of

(270 * 1K) / (270 + 1K), or approximately 212 Ohms.

This leaves us with 15 Volts across the series combination of 212 Ohms and 200 Ohms. The current through this circuit is:

15 / (212 + 200), or 36.4 milliAmps

So how much of this current goes through R5? The voltage at Q3 (Remember, it's saturated and we're considering it to be a short) will be the voltage drop of R4. R4 has all 36.4 milliAmps through it so its voltage drop is:

36.4mA * 200 Ohms = 7.28 Volts

The +15V supply voltage minus this 7.28 Volt drop appears across both R5 and R3. The current through R3 is:

(+15V - 7.28V) / 1000 Ohms = 7.72 milliAmps

This is probably within the specified current limit of Q3's base, but you have to check the data sheet.

Anyway, this circuit kind-of sucks. It really doesn't regulate current, although it does regulate voltage when everything is operating correctly. This looks like a case of exchanging robustness and elegance for expedience.
Nuff said!

Hey! I can't end without proposing at least one solution. The circuit below has one active component added, two resistors with changed values and one resistor has been eliminated. Q4 is in a series pass current limiting configuration. Let's say that the emitter-to-base voltage drop is 0.7 Volts. When 0.010Amps (10 milliamps) flows through R4, the voltage drop across R4 will be approximately 0.7 volts. Anything above this will turn Q4 on. Q4 will conduct through R3 causing the voltage at the base of Q3 to lower. Q3 will then turn off. The value of R3 has been increased because we don't want Q4 to be fighting the output of U2. We could actually have used 4.7kOhms or even more on the circuits above, but what the Hey!

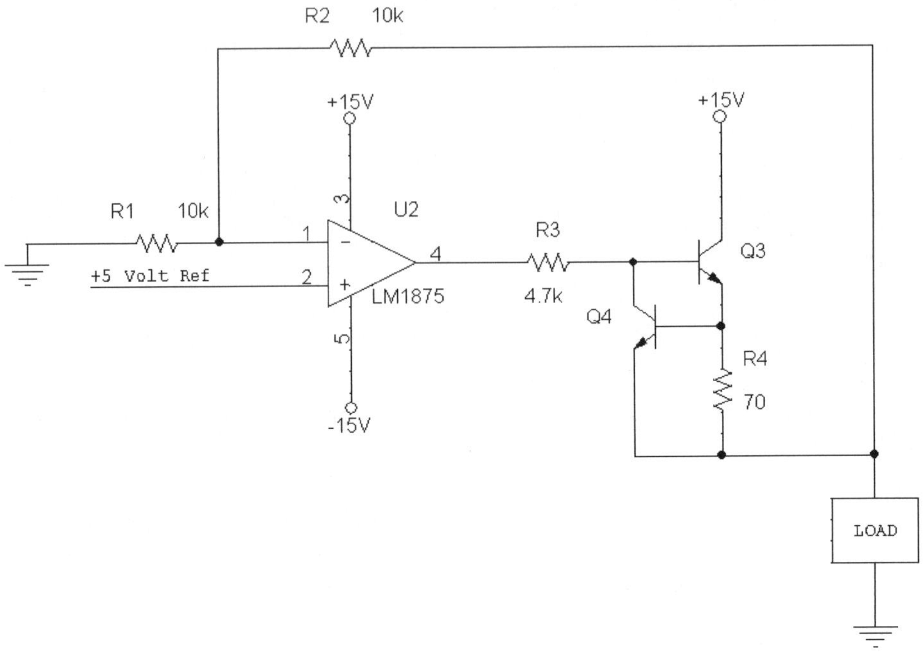

This is not an ideal situation, but if the amount of current you need to regulate is 10mA +/- 1 mA I think we're there. Even with the LOAD shorted to Ground, the current won't be much more than 11 or 12 milliamps (with the load shorted to ground, the current through R2 is zero). Also notice that the collector resistor of Q3 has been removed. It isn't necessary as long as Q3 can handle around 160 milliWatts. Are you worried about the wattage of R4? Using W = EI we get:

W = 0.7 * 0.010 or 7 milliWatts.

I think a 10th watt resistor can handle that. How about changing the circuit so that it becomes a 100mA current limited circuit. How? Change R4 to 7 Ohms. Now the wattage dropped by R4 becomes:

W = 0.7 * 0.100 or 70 milliWatts.

Still under the 10th watt rating. But, in the event of a short, Q3 will have to drop

15V * 100mA = 1.5 Watts. That's a bit stiff.

So, reduce the current to a comfortable level. Hey! It's better than the alternative.

The Bad Boss

Sometimes you just run into a situation that you remember for the rest of your life. Sometimes that isn't exactly what you want, but you have to chalk it up to experience. Face it, working for a living means you have to eat a little bit of crap once in a while. O.K.! O.K! Sometimes you have to eat a lot.

My situation arose when times for engineering jobs weren't that great. In fact, I had just been laid off. In fact, it happened two days after my son was born. Anyway, my mobility was severely limited for several weeks so I had to leave the job-getting in the hands of a 'head hunter'. Usually I like to find my own work, even if it means physically knocking on doors. Anyway, this agent worked hard to get me into a job that sounded like a great thing. My first hint that things might not all be peaches and roses came when the interviewer didn't want to show me where I would be working. Never suppose anything.

I wound up working at a bench with approximately two square feet of area. That, and a hard-backed stool were my life for nearly the next year. That would have been tolerable except for the fact that the guy I was working for was one of the most distasteful persons I have every met. His forte was humiliation with a pinch of degradation, things that would be considered illegal today. Along with this came massive amounts of negativity and absolutely no praise.

The lesson here is; when going for a new job or changing jobs "Be sure you know as much as possible about the work environment, the responsibilities of the job and the person to whom you will be directly reporting."

Even though you may wind up with the greatest boss, things change. The average is about three years before something gets shaken up and you either wind up working for someone new and having to prove yourself all over again or your whole work environment changes. There are many changes along the road called 'working for a living'.

Once again, good luck.

'Cheap and Dirty' Voltage-to-Current and Current-to-Voltage Transistor Circuits

Here are a couple of circuits that can be used for 0-5V to 4-20mA conversion and vice-versa. First, the 4-20mA to 0-5Vcircuit. Assume V+ to be 5 Volts although, it could be a little lower or much higher.

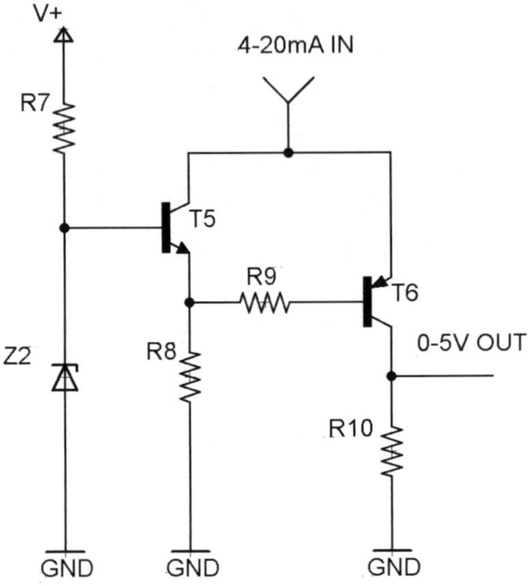

This circuit definitely requires some values as the results are highly dependent upon the component values. So here's a BOM for the circuit.

Bill of Materials

R8	649Ohms	1/8 Watt
R7	2210Ohms	1/8 Watt
R10	316Ohms	1/8 Watt
R9	1kOhm	1/8 Watt
T6	2N3906	PNP
T5	2N2222	NPN
Z2	3.3Volt	Zener

Well, now let's look at what's happening. The voltage at the base of T5 is clamped at 3.3Volts. This makes the voltage at the emitter of T5 go no lower than

3.3V-0.7V or approximately 2.6Volts. With T5's emitter resistor R8 at 649 Ohms, the maximum current through R8 will be 2.6V/649Ohms or 0.004Amps (4mA). At this current, transistor T5 is still in saturation and the Emitter-to-Collector voltage drop will be 0.2V. This voltage is not enough to turn on the emitter-to-base junction of PNP transistor T6, so there is no current in the T6, R10 path. Therefore, the voltage drop across R10 will be 0 Volts. The voltage OUT will be 0 volts.

When the current IN increases above 4mA, the T5 circuit can't supply any more current, so T5's Emitter-to-Collector voltage will increase (T5 will come out of saturation). When this voltage increases to 0.7V, The emitter-to-base junction of T6 will become forward biased and T6 will begin to conduct. There will still be 4mA through the T5 path, but all the rest of the current IN will pass through T6 and R10. As the current IN increases linearly, there will be a linear increase in voltage drop across R10.

When the current IN is 20mA, 4mA passes through T5 and R8 and 16mA pass through T6 and R10. The voltage drop across R10 will be 316 Ohms x 16mA, or 5 Volts. What about R9, the 1KOhm resistor? It's just there to limit the current through the emitter-base diode of T6. This helps to prevent this diode from stealing some of the current that passes through R8, as this might affect the operation of the 4mA circuit (Z2, R7, T5 and R8). So much for that! How about converting 0-5V into 4-20mA on the cheap? Check out this circuit.

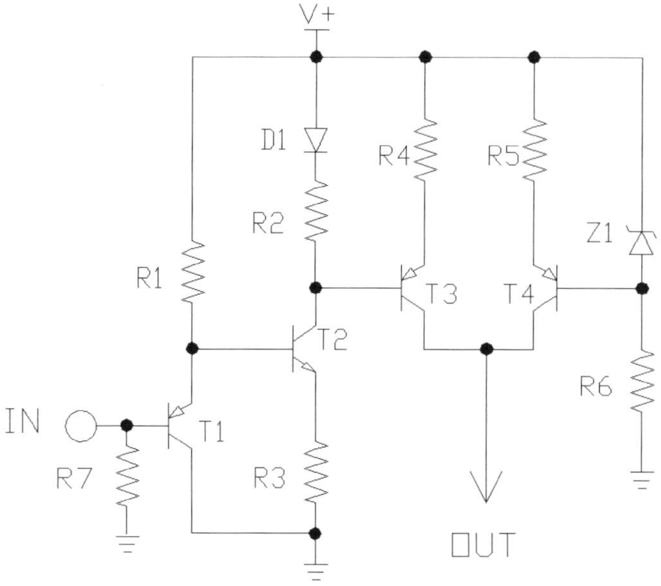

Again, we're going to need a Bill of Materials to fully understand what's going on.

Bill of Materials		
R1	2.2kOhms	1/8 Watt
R2	520Ohms	1/8 Watt
R3	1kOhm	1/8 Watt
R4	162Ohms	1/8 Watt
R5	649Ohms	1/8 Watt
T1,T3,T4	2N2905	PNP
T2	2N2222	NPN
R6	2210Ohms	1/8 Watt
R7	10kOhms	1/8 Watt
Z1	1N5226B	3.3Volt

With 0V at the Input, current through R7 causes T1 to conduct and T2 to be forced not to conduct. With T2 not conducting, T3 will also be forced not to conduct. Well, what about the other half of the story. The Z1, R6 circuit produces a voltage at the cathode of Z1 that is 3.3V below V+. This forward biases the emitter-base junction of PNP transistor T4, forcing T4 to conduct. The voltage at the emitter of T4 is V+ -3.3V +0.7V or approximately 2.6V below V+. This places 2.6V across T4's emitter resistor, R5. The current through R5 will be 2.6V/650 Ohms or 4mA All The Time! So we get the idea that, like the current-to-voltage circuit above, there are two separate circuits contributing to the Output current of this circuit. How about the current contribution from T3? How does it relate to the input voltage? Let's check the current out of T3 with a 5 Volt input. 5 Volts on the base of T1 will produce 5.7V at the emitter. This voltage is present on the base of T2, so 5 Volts will be present on the emitter of NPN transistor T2 and the voltage across R3 will be 5V. The current through R3 will then be; 5V/1kOhm or 5mA. This 5mA passes through R2, a 520 Ohm resistor and through D1, a silicon diode (0.7V drop). The total voltage drop is: V+ -[(5mA x 520 Ohms) + 0.7V] or V+ -3.3V. How interesting! Now we know that V+ must be above a certain voltage. Let's figure out what it is. The drops will include R3, T2, R2 and D1. Assume that T2 is saturated and that its collector-to-emitter voltage drop is 0.2V. The total Voltage dropped is 5V +0.2V +3.3V +0.7V or 9.2V. A little overhead would be nice.

Getting back to T3 and its current contribution, the voltage on the base of T3 is V+ minus 3.3V. So the voltage across R4 is approximately 2.6V and the current through R4 is 2.6V/162Ohms or 16mA. Adding the 4mA through T4 and this 16mA gives us 20mA out with 5V in.

So what else is going on in this circuit? Why don't we need a resistor between the collector of T2 and the base of T3? It's because T3 has an emitter resistor (R4) and this resistance is reflected back into the base (because of the gain or Beta of the transistor).
Actually, the effective resistance between the collector of T2 and the base of T3 is the value of R4 multiplied by the Beta of the transistor. The result is a resistance high enough to make T3's base current minimally significant in our discussion.

How about the linearity of the circuit? The circuit maintains its linearity because the individual active component circuits are always saturated. The transistors are trying as hard as they can, and are not operating in their linear regions because they don't have enough base drive.

Well, I said it was cheap.

Losing the Job at the Interview

I had a few occasions,... well, let's be honest. I had a lot of occasions to interview for jobs. Whenever I left or lost a job I had, on average five to ten interviews before getting another job offer. I can almost remember all of the interviews that didn't work out. Some of these were for the silliest reasons, some were because I didn't know what to expect and some were just circumstances.

About the most memorable interview I had was one during which there was a call for ME! It was my present boss telling me that our work group, including him, was being eliminated and we were all losing our jobs. I wasn't happy with my present job anyway, as evidenced by the fact that I was on this interview, but the call, no matter what was said, so set back the people interviewing me that I never heard from them again. Receiving ANY phone calls during an interview is a No-No!

Another memorable interview was an interview for which I had to travel by plane from Connecticut to Florida. I got there and was shown around during the morning. In the afternoon I had an interview with the General Manager. As we were talking he asked me about a publication I had co-authored. The article was mentioned on my resume and it dealt with electronic flea collars. The bottom line of the article was that the flea collars didn't work. I got a little bit of an argument from the General Manager and I had to defend the findings. When the interview was over I wasn't contacted again. My guess is that he had purchased one or more collars and felt that they had worked great.

During another interview where I had to travel from Connecticut to Georgia, I took along some examples of my designs. As it turns out, this wasn't a good idea. No interviewer wants to get that in-depth. They don't mind you talking about a circuit and maybe scribbling one on a piece of paper in order to better explain it, but reading your stuff is a waste of time. It's the human element that makes or breaks you as a candidate. From then on I only took myself and a couple of copies of my resume. Anyway, that wasn't what chilled the deal on that interview. There was some kind of urinating contest between the person who had invited me to the interview and his boss. His boss was the one who squashed the deal. This relationship was obvious once I got there. Like many other interviews, I wondered why they had wasted their time and mine.

After many of these interviews I actually had felt that I might get the job. But this was short lived when there was no phone call. You can never tell what goes on after the interview. Either someone didn't like something about you or the ideal candidate walked in after you.... There are lots of reasons. You have to get over it and just keep trying.

There were several interviews during which I knew this wasn't the job for me. But I would always try my best to 'get the offer' anyway. This is something I learned from a placement agent, better known as a "Head-hunter". Head-hunters can come in very handy but always remember, they are probably sending out other candidates to interview for the same job for which you are interviewing.

That's about it.

The Schmitt Trigger Input with Bi-Polar Transistors

The Schmitt Trigger is quite a useful circuit. It is an excellent way to minimize the effect of noise on the input to logic circuits and it can be used to set high and low trip points for analog circuitry. It does this by ignoring changes in the input that are above a certain low limit and below a certain high limit. In most logic circuits with Schmitt Trigger inputs, this range is from 1/3rd to 2/3rds of the applied voltage (5 Volts for TTL logic and up to 18 Volts for CMOS logic). Using discretes to create a Schmitt Trigger circuit allows for any applied voltage and any high and low trip points.

The first thing to understand is how the darned Schmitt Trigger works. Let's say that the input voltage is below the low trip point. This means that the output of the Schmitt Trigger circuit will be at a particular state until the input increases above the low trip point and then continues until it gets to the high trip point. When it reaches the high trip point, the output will change states. So the input is now at or above the high trip point. The output will not change state again until the input has moved lower than the high trip point and continued moving lower until it is below the low trip point.

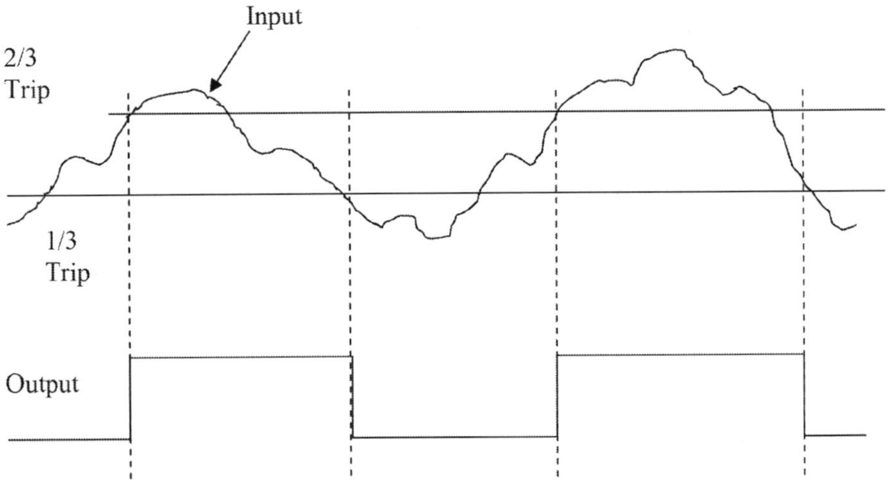

The circuit below shows bi-polar NPN transistors arranged in a Schmitt Trigger configuration. One condition of this circuit is that R3 is lower in value than R2. In order to obtain sufficient hysteresis, the resistance of R2 has to be quite a bit greater than that of R3. The current path consisting of R2, Q1 and R1. This path will have a maximum current when Q1 is saturated (approximately 0.2 volts from emitter to collector). When this occurs, the voltage drop that appears at the top of R1 will be the input voltage (IN) minus the emitter-base voltage drop of Q1 (approximately 0.7 volts). When Q1 is saturated, it effectively shorts out the emitter-base junction of Q2 and therefore keeping Q2 from conducting (off).

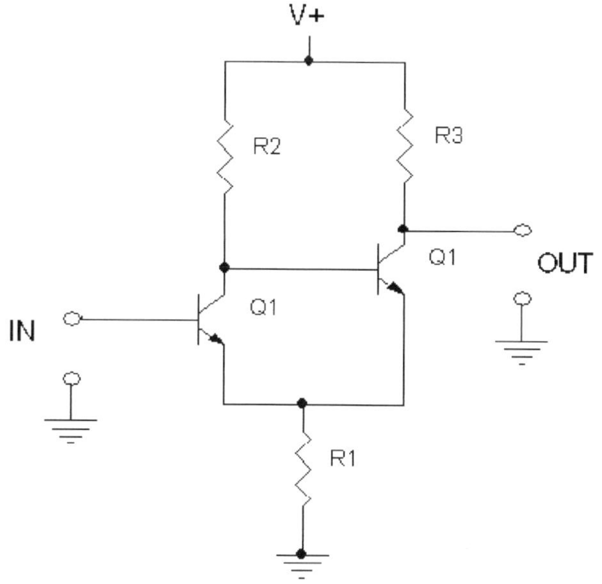

Only when the IN voltage drops below the voltage at the top of R1 (plus the Q1 emitter-to-base voltage drop) Q1 will begin to turn off. As Q1 turns off, it's emitter to collector voltage begins to increase. When the emitter-to-collector voltage across Q1 exceeds 0.7 volts, Q2 will begin to turn on. This happens quite sharply because as soon as Q2 begins to conduct, additional current from the R3, Q2, R1 circuit passes through R1 and causes a greater voltage drop across R1. This decreases the voltage across the emitter-base junction of Q1 and Q1 turns off even faster. So what we wind up with is Q2 fully on and Q1 fully off. The voltage at the top of R1 will have increased because R3 is lower in value than R2 and more current will flow through the R3, Q1, and R1 circuit. More current through R1 means more voltage drop. In order to get Q1 back in the ON condition again, the IN voltage will have to exceed the voltage at the top of R1 plus the emitter-base voltage drop of Q1. Once this happens, Q1 will begin to conduct and as soon as the voltage drop across Q1 is less than 0.7 volts, Q1 will effectively 'short' Q2's emitter-base junction and Q2 will begin to turn off. This change will also happen quite abruptly. In this state, Q1 will remain on until the IN voltage drops below the voltage at the top of R1 (plus the emitter base drop of Q1). The output voltage of this circuit can be taken across R2. We'll examine this a little later. But now let's investigate a particular situation. Lets say that we want a hysteresis of 1/3rd of V+ that has a low threshold of 1/3 V+ and a high threshold of 2/3 V+ and let's make V+ equal to 90 VDC. We will also fix R1's value at 10KOhms.

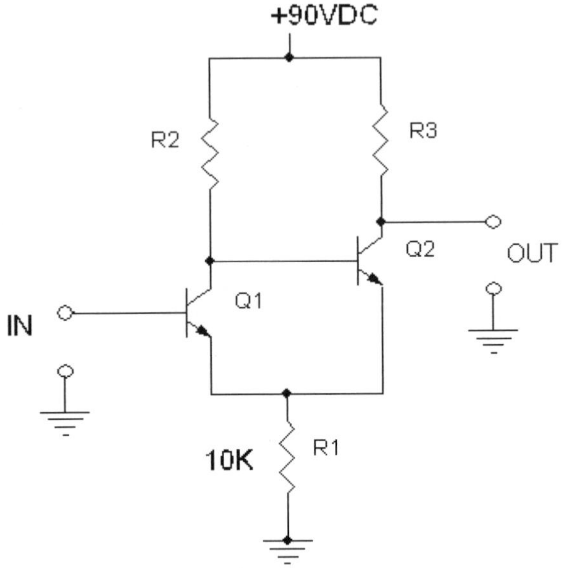

When Q1 is saturated, The current through the R2, Q1, R1 path must produce a voltage drop across R1 such that R1's voltage drop plus 0.7 volts (Q1 emitter-base voltage drop)
Equals an IN voltage of 1/3rd V+ or 30 Volts DC. So 29.3 volts across R1 means that this current is 29.3/10K or 2.93 milliAmps. This same current must flow through R2. The voltage drop across R2 is approximately 90 Volts minus 30 Volts or 60 Volts. The value of R2 is: 60/0.00293 or 20,478 Ohms (approximately 20.5KOhms).

So how about R3? Remember, when Q2 is conducting, the voltage drop across R1 must be almost 2/3 V+ . This keeps Q1 from conducting until the base of Q1 has 2/3 V+ applied to it. In order to get a voltage drop of 2/3 V+, or 60 VDC in our cases, the current through R1 must be: 60/10,000 = 6 milliAmps. This current flows through Q2 and R3 and the resulting voltage drop across R3 is approximately 30 VDC (V+ minus the 60Volt drop across R1). Therefore, the resistance value of R3 is: 30/0.006 or 5000 Ohms.

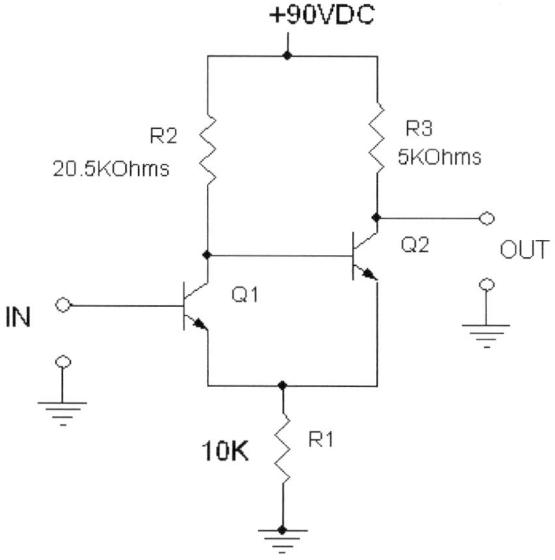

For lower values of V+, the voltage drops of Q1 and Q2 become more of a factor and must be included in the calculations. Notice that the wattage of R3 is less than 0.2 Watts BUT the wattage of R1 is approximately 0.36 Watts ($W = E^2/R$, = 3600/10000 or 0.36 Watts). I promised to cover the output of this circuit back at the beginning. So here goes. Probably the best place to take the output is across R3. Using the values in the above example, when Q2 is on, R3 is at about 30 Volts. When Q2 is off, R3 is close to V+, or 90 Volts.

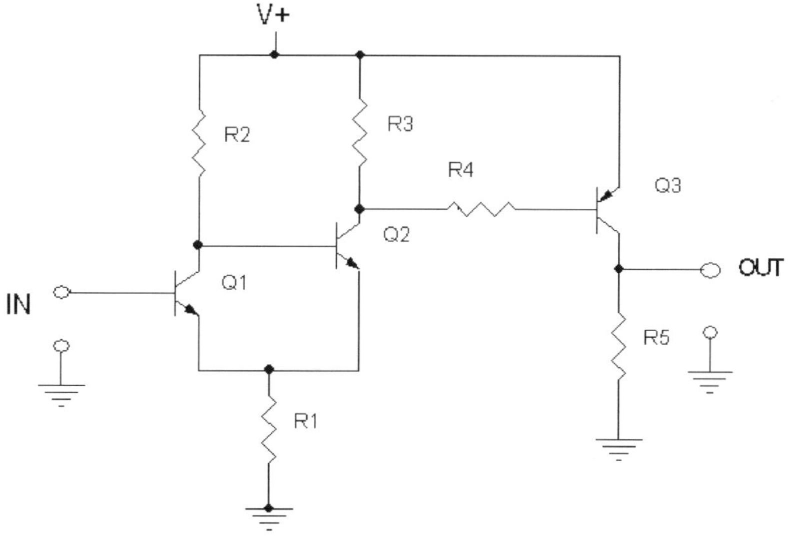

In the figure above, resistor R4 has been added to the collector of Q2. The other end of this resistor goes to the base of transistor Q3, a PNP device. In order for Q3 to conduct, the base voltage must be approximately 0.7 volts below the emitter voltage. This can only happen when Q2 is conducting. The new resistance, R4 limits the amount of current that travels through the base of Q3 and its value will be quite high (100KOhms or greater) so as to produce only a minimal effect on the 2/3rd V+ switching point of the Schmitt Trigger and to limit the base-to-emitter current through Q3. Resistor R5 has also been added. This resistor is located between the collector of Q4 and Ground. This resistor produces a voltage drop only when Q4 is conducting. The top of this resistor will be at either zero volts or at close to +V depending upon the state of the Schmitt Trigger circuit. When the input voltage to Q1 is below 1/3 V+, Q2 will be in conduction. This will turn Q3 on and cause the voltage at the top of R5 to be approximately +V. So when the input to the total circuit is low (below 1/3rd +V) the output will be high (nearly +V). Conversely, when the input is above 2/3rd of +V the output will be at Ground potential. Another way to minimize the affect of the base circuit of Q3 on the operating point of the Schmitt Trigger is to add a resistor between +V and the emitter of Q3 (R6). The figure below illustrates this by the added resistor R6. This doesn't have to be a large resistance because its value will be reflected into the base circuit after being multiplied by the Beta, or gain of the transistor (Beta is normally between 50 and 200 for general purpose transistors).

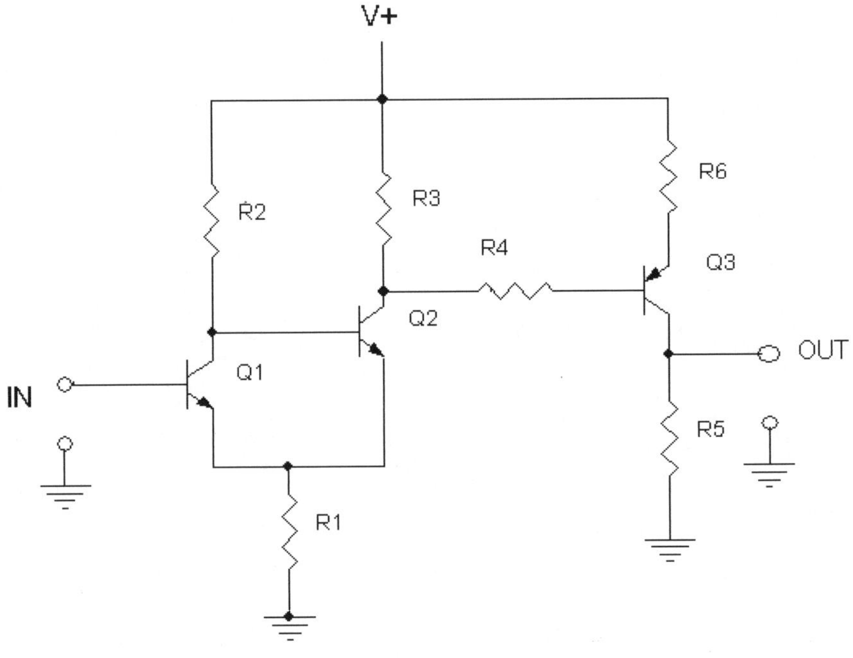

So, with a gain (Beta) of 100, a 2KOhms resistor in the emitter circuit of Q3 will produce an apparent resistance in the base circuit of 200KOhms. The only problem is that the output voltage across R5 will be somewhat lower than +V. If that is not a problem, a resistor in the emitter circuit of Q2 is a good way to go.

That's about it for this circuit. Be careful to note the voltages that the transistors must withstand and use parts that are at least twice this value if possible.

I don't know if you happened to notice, but this circuit and the transistor current-to-voltage converter (4 to 20 mA to 0 – 5 Volts) operate in much the same way. They both have one transistor across the emitter-base junction of another transistor. This enables one transistor to control the conduction of the other, based upon whether or not it is in saturation (full ON).

Just something to think about.

The Good Job

Sometimes things just go right. You have a good interview and you feel comfortable with everyone you met except maybe a few people but you don't have to answer directly to them. The person you are going to work for is someone you feel you can learn from, and also someone you can respect for what they know about some aspect of electronics in which you need improvement. You are sure that you can contribute a lot to the projects on which they are working and they understand this.

Well, cheer up bunky! The average time that this situation lasts is approximately three years, so enjoy it while you can. The reasons why it will change are as follows: 1) The company is sold. 2) The company merges with another company. 3)Your boss is promoted. 4)Your boss leaves for another job. 5)Your boss gets sick. 6)There is a re-organization. 7)YOU are tempted away to another job with promises of more money and more responsibility. 8)The company moves to Afghanistan. 9)Your job is outsourced. My three year average is based upon personal experience.

The worst situation I was ever in involved being in a position that was below what I felt to be my level of knowledge and education, but one in which the money was good. The money was actually a trap. I suffered for almost five years in that position, always feeling I hadn't gotten the responsibility and respect that I felt I had deserved. I know, I know! I should have done something about it, such as taking a lower paying job somewhere else. That's the tough part, especially when you have family obligations and mortgage payments to make.

Here's a fairly common situation that has been present in industry for several decades now. You get a job where they encourage education. So you consistently take courses toward a degree. Once you achieve the degree, the company really doesn't have a position for you. When you were hired they needed a technician. They still need you as a technician. Usually in that situation you are forced to seek another employer just to get ahead. This, I feel is one of the great anomalies of the corporate world.

My best advice is to enjoy the job and the environment while you can. When the situation changes, it's a crap shoot once again.

The Howland Current Pump

The figure below is one configuration of the famous Howland Current Pump. It is used to convert voltage to current, independent of the load. Operation of this circuit is not obvious. It doesn't have the standard operational amplifier configuration which is one input to the op amp of a known, or fixed value. The reason for this is that the output is fed back to both the positive input and the negative input. So there is no single input that is fixed. The same rules apply, that is, the inputs to the op amp always want to be the same. A bit of algebra is required in order to determine how this circuit works.

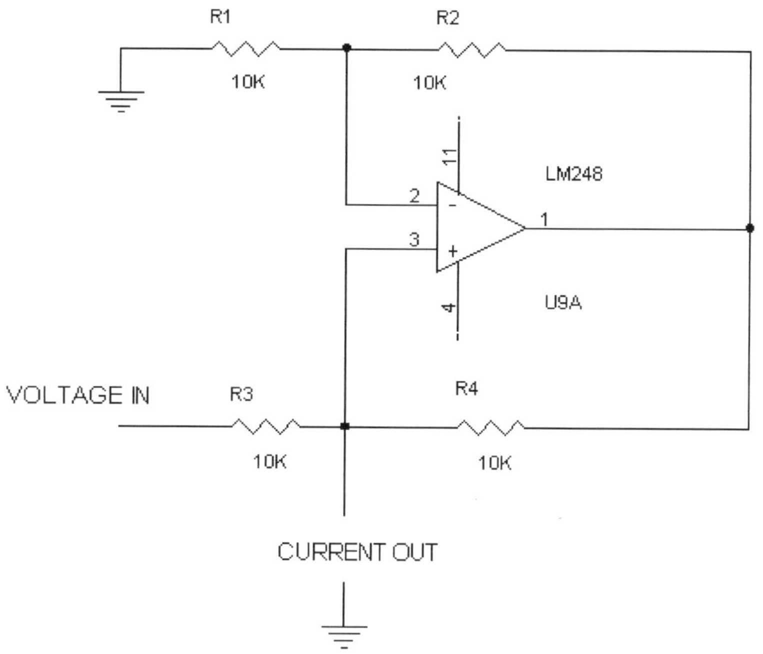

O.K.! What do we know from looking at this circuit? Because of the feedback from the output to the "-" input of the op amp and because R1 and R2 are of equal value, the "-" input will always have ½ the voltage value of the output. The "+" input of the op amp will also have this voltage value (inputs always want to be equal). Notice that R1 and R2 do not have to be 10KOhms. They can be any reasonable value (the higher the better, without being affected by the op-amp input) as long as R1 = R2.

How do we approach the problem? Let's start with what we know. First, the voltage at the "-" input is equal to half of the output voltage (call the voltage at the "-" input Va);

$$Va = \frac{Vout}{2} \quad \text{or} \quad Vout = 2Va$$

Now let's look at some currents. The current through R3 is;

$$\frac{\text{VOLTAGE IN} - Va}{R3} = i1$$

Call the output of the op amp "Vout". The current through R4 is;

$$\frac{Vout - Va}{R4} = i2$$

(Notice that we are assuming the current through R4 is flowing from Vout toward Va).

We know that the CURRENT OUT is equal to i1 plus i2 (the "+" input to the op amp doesn't draw any current). So we get the formulas;

$$\text{CURRENT OUT} = i1 + i2$$

Substituting for i1 and i2:

$$\text{CURRENT OUT} = \frac{\text{VOLTAGE IN} - Va}{R3} + \frac{Vout - Va}{R4}$$

Since R3 and R4 are equal we can just call them "R".

Remember that Va is equal to $\frac{Vout}{2}$

Substituting this into the above equation yields;

$$\text{CURRENT OUT} = \frac{\text{VOLTAGE IN} - \frac{Vout}{2}}{R} + \frac{Vout - \frac{Vout}{2}}{R}$$

The '$Vout - \frac{Vout}{2}$' term becomes just $\frac{Vout}{2}$

Since both terms are divided by R, simply add the $+\frac{Vout}{2}$ to the $-\frac{Vout}{2}$ terms,

leaving noting but the "VOLTAGE IN" term, divided by R.

The result is;

$$\text{CURRENT OUT} = \frac{\text{VOLTAGE IN}}{R}$$

Holy cow! What a simple relationship. But is it true? Let's substitute real values again and see what happens.

Example:
In the above circuit let Vin equal 10 Volts. Then the Current Out will be 10V divided by 10KOhms or 1 milliAmp.
Take a situation where the load resistance is greater than zero Ohms and a value of 1Kohm for the load resistor.

The current through the 1Kohm load resistor must be 1mA (as long as the circuit is not in saturation). This means that the voltage drop across the load resistor is 1 Volt. Therefore, the voltage at pin 3 of the LM248 is also 1 Volt. Since the voltage on pin 2 is the same as the voltage on pin 3, the voltage on pin 2 will be ½ Vout. So Vout will be equal to 2 Volts. Now let's figure out the current through R3 and then R4;

$$\text{Current through R3} = \frac{\text{VOLTAGE IN} - \text{V(pin 2)}}{\text{10KOhms}} = \frac{\text{10Volts} - \text{1 Volt}}{\text{10KOhms}} = \text{0.9mA}$$

$$\text{Current through R4} = \frac{\text{Vout} - \text{Va}}{\text{10KOhms}} = \frac{\text{2Volts} - \text{1 Volt}}{\text{10KOhms}} = \text{0.1mA}$$

The sum of these current is 1mA, the desired current.

So the darned thing works! At least in this instance. You may wish to try a few more variations just to be sure. One thing to note is that for any particular input, there is no particular output from the op amp. When load resistance changes, the current through the load remains constant (or directly related to the input voltage). The only thing that can vary in order to accommodate these load changes is the op amp output voltage.

The simpler method:
Knowing what this circuit does leads to one of the easiest circuit solutions yet. The "current out" has to be the same, whatever the load. So let's consider the load to be zero Ohms. Simply ground the junction of R3, R4 and pin3 of the LM248. Now, the "voltage in" is applied entirely across R3. Pin 3 of the LM248 is held firmly at ground. Therefore pin 2 of the LM248 will also try to be at ground potential. The only way this can occur is if there is no current through the resistor R1 which is connected to ground. No current through R1 means no current through R2 which means no voltage (ground potential) at the output of the

LM248 and thus no contribution to the output current from the LM248. All of the load current will pass through R3. This leads to the formula:

$$\text{Current Out} = \frac{\text{Voltage In}}{\text{R3}}$$

Look Familiar?

The only short-coming of this circuit is that the output currents are quite small. Our example used 10KOhm resistors but if we were to use 1KOhm resistors, the CURRENT OUT would try to be 10 milliamps with 10 Volts in. With a 1KOhm load the voltage drop across the load would be 10 Volts. Ten volts in (VOLTAGE IN) and ten volts for Va mean zero current through R3. This means that all of the 10 milliamps has to come from the output of the op amp. But the op amp would have to produce 20 volts in order for this to happen. This is well above most op amp positive power supply voltages. Then too, most standard op amps can't source or sink 10 milliamps, but that's another story.

Being limited to low currents is a serious circuit drawback for the Howland current pump. But help is on the way! Check out the section on the Improved Howland Current Pump.

The Older Working Engineer

I didn't notice it until a little later in my working career, but as I grew older it became more evident that youth is preferred over experience, especially in the engineering profession. Several years ago an Electrical Engineering Department was managed by an engineer who had probably worked his way into this management role while at the same company. More and more, the role of management was passed to employees or outside hires with MBAs rather than to engineers with BSEEs or MSEEs. MBAs are trained for managing just about any department, or so it is believed.

At this point I will contain myself, although I do have my own feelings about the reasons for this. O.K., O.K.! Here it is. Corporate management, as a group has the combined intelligence of a fifth grader. For this reason they are very susceptible to what I call the "Band-Wagon" effect. It's easy for a corporation to jump on the Band-wagon, thus causing other corporations to do the same so as not to be left out, just like a third grader. Need a few examples? How about robotics, just-in-time systems, ISO certification, IT departments, and on.. and on. There! I've said it. Saying it doesn't fix anything, it just supplies reasons, and engineers do need reasons.

Anyway, the engineer just became another cog in the wheels of the corporate machine. He was replaceable and expendable. It wasn't long before those who calculated the corporate 'bottom line' began to realize that hiring engineers that were between entry level and 3 years of experience would save the company a lot of money. These engineers would be familiar with the new technologies not to mention, they came cheap. Then too, the corporation could train the young engineers the way they wanted them to be. Fat chance!

So where did this leave the older engineers? First, their jobs were in jeopardy. They had better have kept up with the latest technologies or they might as well go into sales. For those that did keep up technologically, they were still viewed as 'old' and there was not much they could do about that. Their only ace-in-the-hole was that they also knew the old technologies upon which most of these corporations were built. But in many cases, this didn't figure into the 'bottom line' and they were eased out. Many corporations lost a lot of legacy knowledge when an older engineer was retired or 'let go'. In several cases, these old engineers had to be called back from retirement in order to make stuff work or keep it working.

But heaven help the older engineer who is suddenly in the job market. It doesn't matter that they have a vast amount of experience, that they are diligent and reliable workers, that they take pride in what they make and that they are grateful for the opportunity to work. They are at a serious disadvantage when it comes to getting a job. I can only wish them luck.

The Improved Howland Current Pump

There is a circuit that at least helps the op amp output voltage from trying to exceed the op amp supply voltage (even though the op amp is still limited in the amount of current it can source and sink). This circuit is known as the 'Improved Howland Current Pump'.

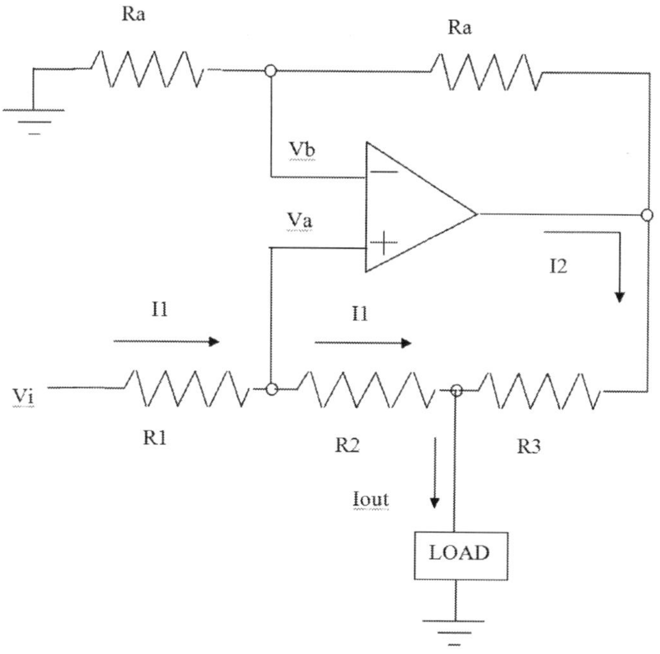

With the addition of one resistor, R3, the output voltage requirement of the LM248 is reduced significantly from that of the original Howland Current Pump circuit.

Using the schematic above, a determination of the General Solution for this circuit can be made:

Let's use the same trick that we learned while examining the Howland Current Pump. Since this is a current source it should not matter what the load is so, **Make the Load be zero Ohms** that is, ground the output.

With the output (junction of R2 and R3) grounded: $I1 = \frac{Vin}{R1 + R2}$

The voltage at Va is; $Va = I1 * R2$

The voltage at Vb is the same as at Va, therefore Vout must equal 2Va.

The current I2 that passes through R3 is: $I2 = \frac{Vout}{R3}$

Since Iout = I1 + I2, $Iout = \frac{Vin}{R1 + R2} + \frac{Vout}{R3}$

Another way of expressing Vout is to say that $Vout = 2Va, = 2 * I1 * R2$

and substituting $\frac{Vin}{R1 + R2}$ for I1 in the above equation, $Vout = \frac{2 * Vin * R2}{R1 + R2}$

From the equation;

$Iout = \frac{Vin}{R1 + R2} + \frac{Vout}{R3}$ we now get; $Iout = \frac{Vin}{R1 + R2} + \frac{\frac{2 * Vin * R2}{R1 + R2}}{R3}$

Multiplying the denominator of the second term by R3 we get:

$$Iout = \frac{Vin}{R1 + R2} + \frac{2 * Vin * R2}{R3 (R1 + R2)}$$

Factoring out $\frac{Vin}{R1 + R2}$ from each term we get $Iout = \frac{Vin}{R1 + R2} \left(1 + \frac{2 * R2}{R3}\right)$

This is the General form of the Vin/Iout transfer function of the Improved Howland Current Pump.

Let's test it out.
If you're convinced that grounding the output is totally valid then we'll use this configuration once again. If you're not convinced, then too bad! we're grounding it in this example.

Given the situation in which R1 = R2 + R3

Select some values for R1, R2 and R3 based upon this relationship.

If R1 = 10KOhms and R2 = 9.9KOhms then R3 must equal 100 Ohms.

For simplicity, let Vin equal 10 Volts

Remember, Ra can be any value (preferably a high resistance value) as long as it is not affected (loaded) by input Vb of the op-amp. 100KOhms sounds good to me. So, what is the current contributed by the R1-R2 path? It will be the input voltage (10V) divided by the sum of R1 and R2, or 19.9KOhms. This current is approximately 5.025 x 10e-4 or 502.5 micro-Amps.

The voltage drop across R2 will be 9.9KOhms multiplied by 502.5 micro-Amps or approximately 4.975 Volts. This will also be the voltage value at Va, the input to the op-amp. Vb will also be at this potential. This can only happen if Vout is 2 * Vb. So Vout is 2 * 4.975V or approximately 9.95 Volts.

This 9.95 Volts is also across R3, the other end of which is to ground. Therefore the current contribution of R3 is approx. 0.0995 Amps. The total current through the load is the 502.5 micro-Amps and the 0.0995Amps or approximately 0.1A (10mA).

So, we have achieved a 100mA output without saturating the voltage output of the op-amp (less than 10V out) using this configuration. I realize that this is some special op-amp because that's a lot of current for most op-amps.

HEY! Notice one more thing. The output voltage is directly proportional to the numeric value of the output current. 10 Volts in, 10mA out. This makes for convenience in design.

Anyway, does our GENERAL FORMULA work in this case?

$$Iout = \frac{Vin}{R1 + R2}\left(1 + \frac{2 * R2}{R3}\right), \quad = \quad 0.1 \text{ Amps, or } 100mA$$

Simplifying the General Form of the Equation:

The last example used a specific relationship between the values of R1, R2 and R3. If we do a little mathematical hocus-pocus we may be able to tailor the General Form of the Improved Howland Current Pump equation to this specific situation.

Here goes!

$$Iout = \frac{Vin}{R1 + R2}\left(1 + \frac{2 * R2}{R3}\right)$$

Expand the expression:

$$= \frac{Vin}{R1 + R2} + \frac{2 * R2 * Vin}{R3\,(R1 + R2)}$$

Multiply the top and bottom of the first term by R3 and combine the two terms:

$$= \frac{R3 * Vin + 2 * R2 * Vin}{R3\,(R1 + R2)}$$

Remembering that this special case requires that R2 + R3 equals R1, substitute 'R1 minus R2' for R3 in the numerator only:

$$= \frac{R1 * Vin - R2 * Vin + 2 * R2 * Vin}{R3\,(R1 + R2)}$$

Simplifying the numerator;

$$= \frac{R1 * Vin + R2 * Vin}{R3\,(R1 + R2)}$$

Combining numerator terms;

$$= \frac{Vin\,(R1 + R2)}{R3\,(R1 + R2)}$$

Canceling terms:

$$Iout = \frac{Vin}{R3}$$

How about that! What a simple transfer function. Remember, it's only for this special case.

Here's something to help minimize unique resistor values:
Suppose you don't care about the convenience of calculating voltage-to-current but that you want to minimize the number of unique resistor values in the circuit. To do this, let's set R1 equal to R2. We can also set the Ra's equal to R1 as long as they don't take too much current away from the op-amp output.

We'll do the mathematical gymnastics first and try to get a new formula for this situation.

One more time, with feeling.

$$Iout = \frac{Vin}{R1 + R2}\left(1 + \frac{2 * R2}{R3}\right)$$

Expand the expression:

$$= \frac{Vin}{R1 + R2} + \frac{2 * R2 * Vin}{R3\ (R1 + R2)}$$

Since R1 = R2, substitute R1 for R2:

$$= \frac{Vin}{2 * R1} + \frac{2 * R1 * Vin}{R3\ (\ 2 * R1)}$$

Cancel:

$$= \frac{Vin}{2 * R1} + \frac{Vin}{R3}$$

Factor out a Vin term:

$$Iout = Vin\left(\frac{1}{2 * R1} + \frac{1}{R3}\right)$$

That's it. So if you want to find the best common resistor values you'll have to solve for R3, set up a spreadsheet or use MathCad and a list of common values.

As always, good luck.

Software Guy "Just the spec's please, just the specs!"

This guy I knew, a software guy named Slam, had about the best attitude I can think of when it came to interfacing with hardware guys. Actually, he was quite a coarse character, but I still admire his approach. When you wanted him to solve some problem for you he would just say "what kind of inputs do I have, what kind of outputs do you want and when".

That kind of sums it up. He didn't care about D/A speed, how long a switch contact was there, whether or not it needed debouncing, he would do whatever you spelled out. He wanted a SPEC. Imagine, someone wanting a specification before they did the work. People didn't care for his request because it made them think about what they really wanted. The hardware guys were in the habit of trying to educate the software guys about how the hardware prepared the signals for them and here was someone who drew the line at the very inputs and outputs of the computing domain. Actually, if the hardware engineers did a good job of making these specs, all Slam had to do was sit down and code. And why not? That's what he did best.

Apparently, this is the way things are going anyway. If you plan on sending your software effort out-of-house, you'll definitely have to go through this exercise and you'll have to be darned sure that you have those specs right. Otherwise it's the company's money and your back-side. This just points out one more lesson that I have learned during my career. Read the specs but, if you're the guy writing the specs, be sure that you have covered all the bases, just as you would want the spec writer to do.

Happy documenting.

VHDL State Machine Development, Conventions and an Example using CPLDs

So what is a state machine? Well, it is something requiring two or more states or decisions. There can be a change from one state to another state whenever one of the decision elements changes. This change might be an input to the state from a switch or timing element.

There are different types of State Machines. The one used within this document is the "One Hot" state machine, meaning that only one state is active (Hot) at any given time. In CPLD hardware, each state of a "One Hot" state machine would be handled by a clocked latch. A state machine requires a 'clock'. The clock transitions control how fast the state machine decisions are made. In our example, the frequency of this decision making clock must be much higher (at least two orders of magnitude) than the resulting outputs. This will prevent detection of variations in the length of the output signals. Clocks used as CPLD clocks can have frequencies on the order of tens of megahertz. When dealing with audio frequencies, as in this example, the clock frequency can easily be three or four orders of magnitude greater than the resulting output frequency

What we want to do is to create a piece of hardware using a CPLD, which can produce the signals necessary to generate Morse Code (good old Samuel F.B. Morse). You may remember that Morse code is a series of spaced short and long signals, usually generated as tones (except for the old style telegraph, where the length of the signal was the time between clicks).

Anyway, We'll control the generation of the 'dits' (short signals) and 'dahs' (long signals) using two normally-open, momentary switches. One will be for making 'dits', the other for making 'dahs'. As long as the 'dit' switch is closed, 'dits' will be produced. The same applies for the 'dahs'. One condition of these external switches is that both switches can never be closed at the same time. This is normally prevented mechanically by not allowing the other switch to be closed when one is already closed.. Each of these switches will become an input to the CPLD and to our state machine. The only output from the CPLD and state machine will be the spaced 'dits' and 'dahs'. The output pin of the CPLD will be at a high level whenever a 'dit' or 'dah' is in progress and low for all other times. When a switch is closed and a 'dit' or 'dah' is started, even though the switch is opened before the 'dit' or 'dah' has completed, the 'dit' or 'dah' will continue to be generated until it is complete. The ratio of 'dah' time to 'dit' time is 3 to 1. This output is not the actual tone normally associated with Morse code, but is rather the gating pulse for a tone. Additional hardware will be required in order to generate tone signals.

Between the 'dits' and 'dahs' there must be a specific period of time. We will call this time period the 'space'. When a 'dit' or 'dah' is complete, there must always be a 'space' just in case the 'dit' switch or 'dah' switch is closed right away and, in the event that a switch is continuously depressed, to make evenly timed spaces between each of the 'dits' or 'dahs', whichever switch happens to be depressed. . If no key is depressed, the 'space' will simply time out. Since there is no memory used in this example, after the 'space' times out and no switches are depressed the output will not be asserted (go 'low' in our case) and the system will await closure of the next switch. During the 'space' time period, the output of the CPLD will be a low level. The time of a 'space' will be the same as the time of a 'dit'.

O.K., so much for the system inputs and outputs. What about the insides of the CPLD, the state machine itself. Well, before we can start right in on generating state machine code we have to do a little thinking about the problem. What are the various conditions that can exist?

First: There can be a situation where there are no active inputs (neither the 'dit' key nor the 'dah' key depressed) and any 'space' time periods have all timed out.

Second: The 'dit' key can be depressed. This can occur at any time, i.e. when a 'dit' is in progress, when a 'dah' is in progress, when a 'space' is in progress or when nothing is in progress.

Third: The 'dah' key can be depressed. The conditions for this are the same as for the 'dit' key.

Fourth: The system can be in 'space' mode. In this mode, the 'dit' switch can be closed or open and the 'dah' switch can be closed or open but the output from the CPLD will remain low.

Everything has to start somewhere and this is true for the state machine also. So, let's start when there are no active inputs and call this the IDLE state. The next condition, or state will be entered when the 'dit' switch has been depressed. Call this the DIT state. Notice that I say <u>has been</u> depressed. When in the DIT state, the 'dit' key may or may not be depressed but it had to have been depressed in order to get us here. Likewise make the 'dah' switch be the DAH state.

There is only one condition left and that is the condition during a 'space' period. We stated this earlier, but after the 'space' period is entered (by the completion of a 'dit' or 'dah') it stays in this state until the 'space' has timed out, no matter what the inputs do. This becomes the SPACE state.

A pattern is forming here. All states are entered by some condition but the condition doesn't necessarily have to persist in order to remain in that state. Sounds like a latch to me! And that is actually what a state is. It's a latch that is set under a certain set of conditions and reset under another set of conditions.

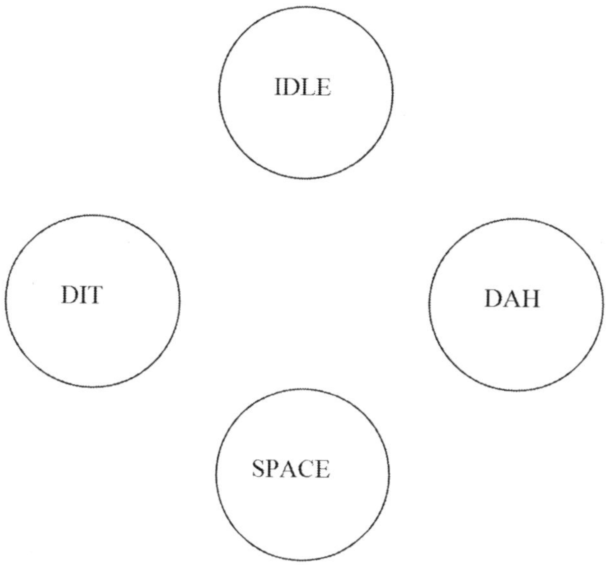

Now that we have defined four states let's examined how they might be connected. What better place to start than in the IDLE state. In order to enter this state we may use something like a built-in power-up reset condition or we could add another input to the CPLD that indicates when power has been applied. We can then add instructions to our program to force the IDLE state when this signal is present. Anyway, we will get into the IDLE state somehow. In this state we'll examine the inputs to the CPLD and either act upon them or stay in this state.

After we're in the IDLE state what can happen? Better yet, what can't happen? Well, to get into the SPACE state either the DIT state or the DAH state must have just completed. But we're in the IDLE state, so the we can't go to the SPACE state from here. Fine, but we can go to either the DIT state or the DAH state depending upon the condition of the CPLD inputs. Then, when either the DIT or the DAH states are satisfied, the next state after them is the SPACE state. Notice that the state following the DIT or DAH states *must* be the SPACE state. Whether or not switches are closed, or have been closed they must go to the SPACE state. When the SPACE state is satisfied the path will be back to the IDLE state where the inputs will once again begin to be examined, starting on the next low-to-high transition of the clock.

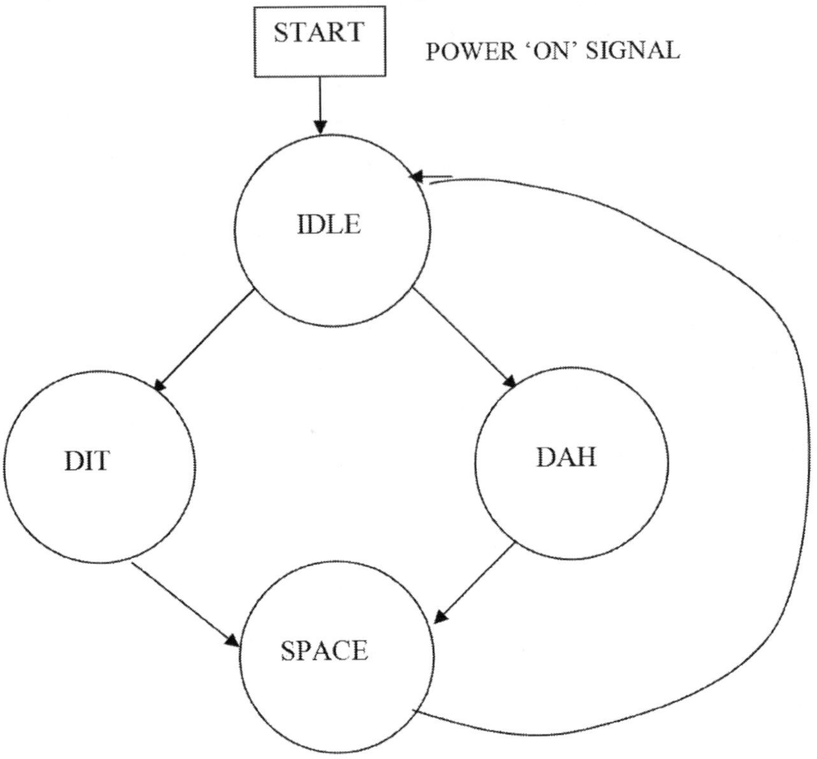

So how do we control the lengths of the 'dits' and 'dahs'? Well, since we have a clock, how about using counters. We'll need counters to count how long we remain in the DIT, DAH and SPACE states. The length of these counters will be determined by the clock frequency and how long we want to stay in any particular state. The counter is incremented with each clock pulse. On our diagram we indicate that a counter is being used by means of a line that comes out of a state and loops back into it. A box next to the loop-back path specifies that it is a counter. To the left of this box is the letter 'T'. This signifies that the counter is being used as a timer. Contained within the box is the instruction that causes the counter to increment, for example; 'counterA = counterA + 1'. Anything contained within a box is considered an 'action' or 'assertion'.

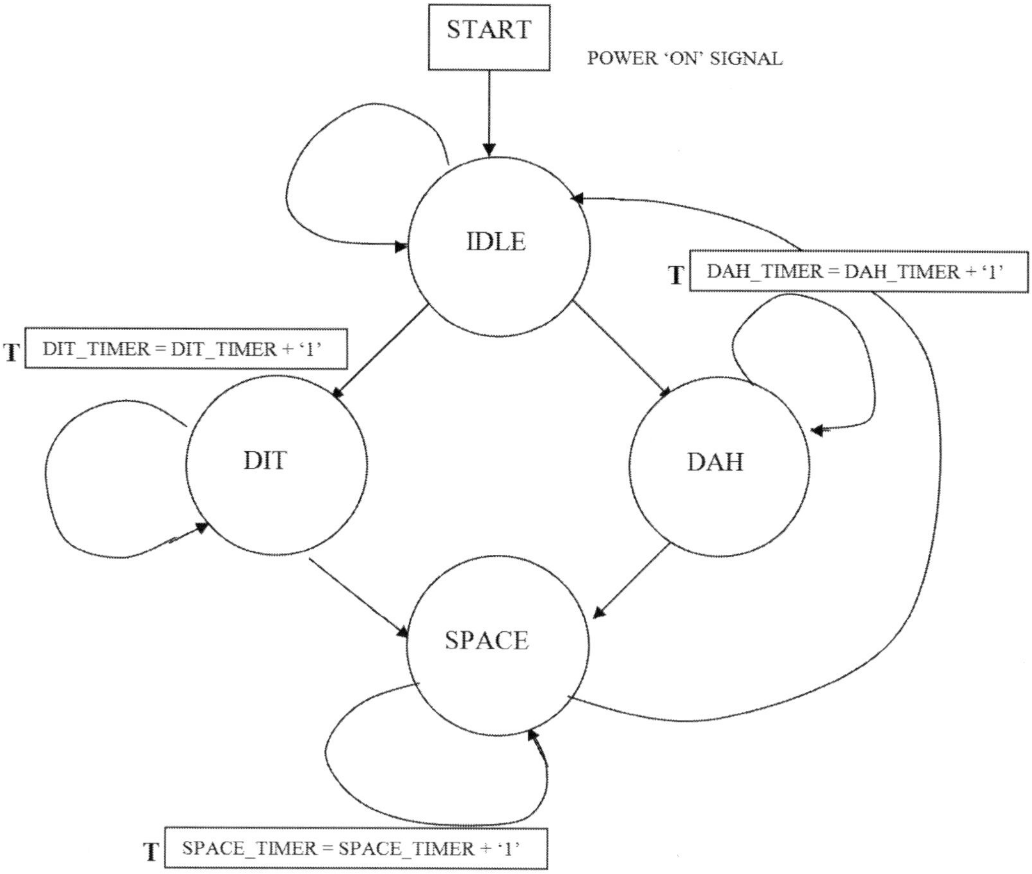

Notice that the IDLE state has a loop-back line but no counter associated with it. This simply means that the machine continues to loop in this state until one of the exit criteria is met, that is the 'dah key' switch or the 'dit key' switch has been closed. During each state the output of the CPLD will be either high or low. This is another action or assertion box that can be added to each state of the diagram.

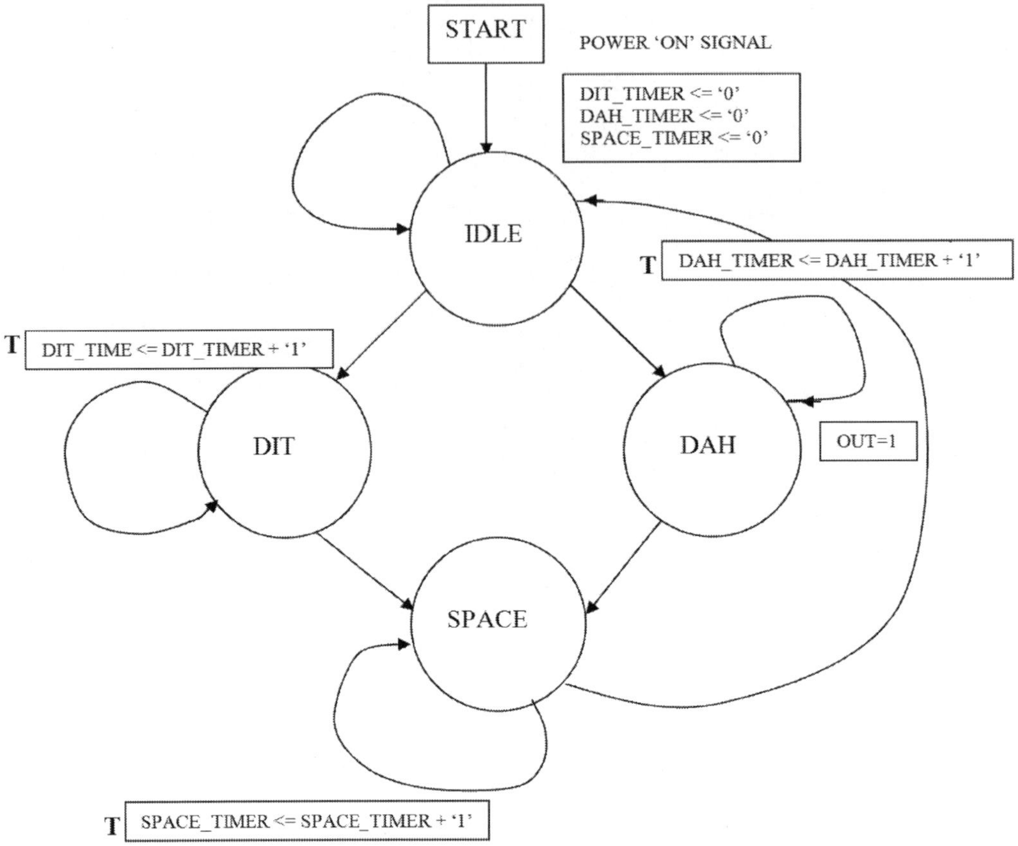

Remember that the box associated with counters had a 'T' next to it? Boxes with letters or arrows next to them indicate a condition of the state machine itself. That means a dependence upon the clock. However, the boxes that don't have anything next to them indicate a combinatorial action (just logic). The combinatorial action is usually based upon one or more of the results of the state machine. For example, the output of the CPLD will be high when in the DIT or DAH states. The boxes (without letters or symbols) that are next to these states designate that combinatorial logic is being used to determine the output. This combinatorial logic may be a two-input OR gate with one input connected to the output of the DIT state latch and the other input connected to the output of the DAH state latch and the output going to the CPLD output pin. That's it. It is not part of the state machine but uses the state machine logic to make a decision. This logic doesn't depend upon the clock. It's just combinatorial, popcorn logic.

Information regarding the condition(s) for moving from one state to another can also be added to the diagram. This information does not appear in a box and is usually in the form of simple statements, such as 'dah_key' which would indicate that the 'dah' switch had been closed. In our example the 'dah_key' statement is associated with the line connecting the IDLE state to the DAH state and is the condition for getting from one state to the other. There are always conditions for getting from any one state to another state. These condition statements are usually referred to as GUARDS (don't ask me why).

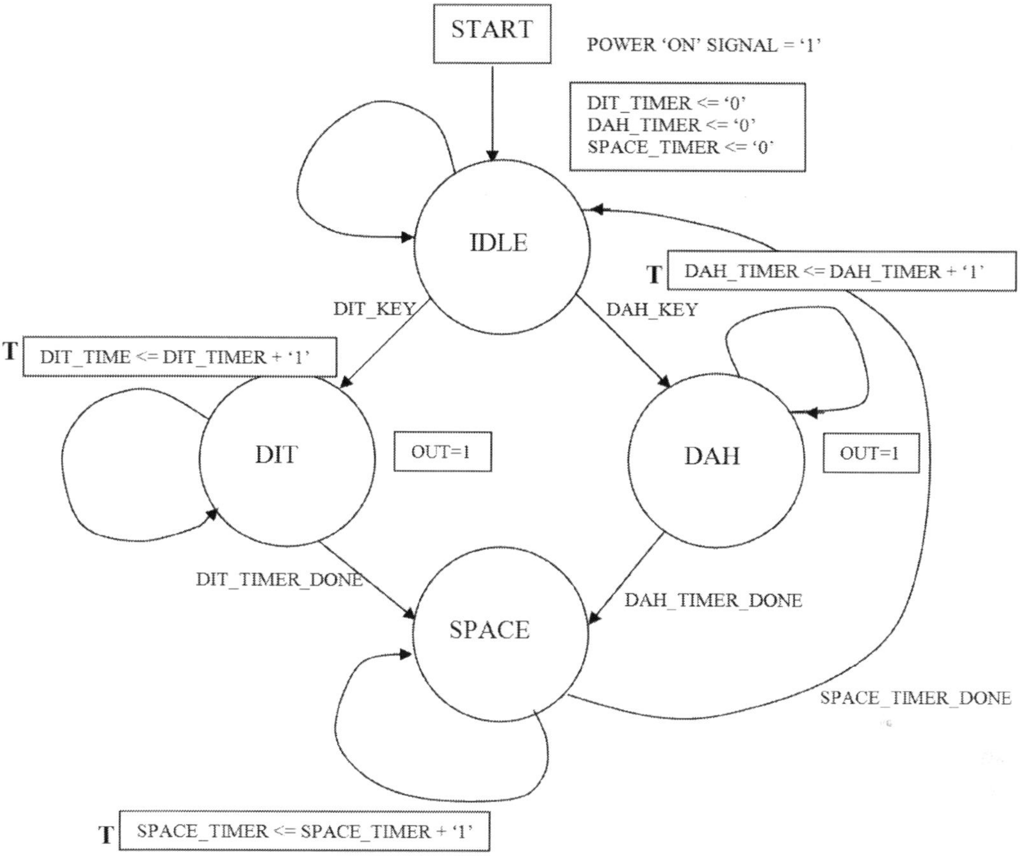

I think it is time to begin coding some VHDL using this bubble diagram.

The complete VHDL program appears at the end of this section, before the section on generating a hardware representation of VHDL code.

There are many good reference books dealing with VHDL. This description is only meant to present the flow of the code in one specific example.
Let's break up the VHDL program into littler pieces.
The first action in the VHDL code is to define the 'include' libraries. These are standard libraries of logic and arithmetic operations that are used by the VHDL compiler to interpret the VHDL code.

In our example the libraries are:

```
        IEEE.std_logic_1164.all
        IEEE.std_logic_arith.all
    and IEEE.std_logic_unsigned.all
```

These are the most common and are usually required for any VHDL program. The next step is to generate the 'entity" section of the code. This section allows the compiler to recognize and determine the type of inputs and outputs to the CPLD. In our example the name of the entity section is 'morse_keyer'. Here is what the beginning of the 'entity' section looks like:

```
entity morse_keyer is port
    (clk:      in std_logic;
     dah_in:   in std_logic;
     dit_in:   in std_logic;
     output:   out std_logic;
    );
```

The format of these instructions must be closely followed so that the CPLD compiler won't throw up. The important things are: the entity statement that starts with the word 'entity' followed by some name (in our case, 'morse_keyer') and ending with 'is port'. In our example, everything after the '(', and before the ');' is then evaluated as an input or an output.

Within this first section of the CPLD code there are certain attributes that are defined. These attributes, in conjunction with 'entity' declarations, relate the CPLD code and it's variables to the actual hardware (CPLD part). One of these attributes is the name of the CPLD part that will be programmed. This can vary from manufacturer to manufacturer and from manufacturer's compiler to manufacturer's compiler. Another attribute relates the signal names in the CPLD code to the actual pins of the CPLD. In our example, these statements are:

1) attribute part_name of morse_keyer: entity is "Cy37256P208-83ac";
2) attribute pin_numbers of morse_keyer: entity is "clk:4 output:73 dah_in:74
3) dit_in:75 ";

The first attribute tells us that a Cypress Semiconductor part will be the target of this code. The second attribute tells us that the 'clock' will be connected to pin 4, that the output will be on pin 73, that the 'dah' switch will be connected to in 74 and called 'dah_in' in the program. The 'dit' switch will be connected to pin 75 and called 'dit_in' in the program.

The next statement in the code ends the entity 'morse_keyer'.

```
end morse_keyer;
```

The next section of the VHDL program is known as the 'architecture' section. It is linked to the 'entity' section by declaring the name of the entity. The architecture has its own name. The architecture section is the last section in the program. Within this section the signals used in the state machine are declared, the state machine itself is generated and any other synchronous or combinatorial code can be found.

Here is the statement that links the architecture section to the entity section, declares the four states of our bubble diagram and tells the compiler that a state machine is to be generated. The architecture section has its own name. In this example it is 'morse_keyer_arch', but it can be almost any name.

```
architecture morse_keyer_arch of morse_keyer is type keyer_states is (idle, dit, dah, space);
```

The next group of statements defines signals to be used in the program. The first signal, 's_current_state' is used by the compiler when the state machine is being defined. Also, in our example there are single variables and counter variables. Signals are not directly related to CPLD pins, but must be related within the code, that is, an entity element such as 'output' (that is directly associated with a CPLD pin) can be passed a value from signal 's_out' (a signal within the program).

```
signal s_current_state:    keyer_states;
signal s_out:              STD_LOGIC;
signal s_dit_counter:      STD_LOGIC_VECTOR (9 DOWNTO 0);
signal s_space_counter:    STD_LOGIC_VECTOR (9 DOWNTO 0);
signal s_dah_counter:      STD_LOGIC_VECTOR (11 DOWNTO 0);
signal s_dah_key:          STD_LOGIC;
signal s_dit_key:          STD_LOGIC;
signal s_dah done:         STD_LOGIC;
signal s_dit_sync:         STD_LOGIC;
signal s_dah_sync:         STD_LOGIC;
signal s_dit_done:         STD_LOGIC;
signal s_space_done:       STD_LOGIC;
```

The next statement signals the compiler that this is the beginning of the state machine.

Begin

Next is the state machine itself. The following statement begins the state machine;

state_machine: process (clk) begin

The next statement makes the beginning of the process dependent upon a low-to-high transition of the clock;

if (clk'event and clk = '1') then

The following statement is a case statement that is used to determine the appropriate state.

case s_current_state is

One of the states is the IDLE state. I say one of the states rather than 'the first state' because the order in which things are listed in the listing are unimportant to the state machine. Just as it would be in hardware, every condition of every state is looked at simultaneously at the transition of the clock.

when idle =>

The above statement indicates that what is to follow occurs in the IDLE state. From our state diagram, all of the bit signals and counters are reset in the IDLE state. The 's_out' signal is an internal signal used by the program that will ultimately control the CPLD 'output' pin, pin 73. The 's_out' signal is set to '0' or low in the IDLE state. The conditions for leaving the IDLE state are listed in the 'if' and 'elsif' statements. If the 'dit' or 'dah' switches are closed the IDLE state is left. This is the only state in our state machine that looks at the 'dit' and 'dah' switches. If the conditions of the 'if' or 'elsif' program statements are not satisfied, the state machine remains in the IDLE state as indicated by the instruction 'else s_current_state <= idle;'.

Here is the IDLE state code:

```
case s_current_state is
   when idle =>
      s_out <= '0';
      s_dit_counter <= (others => '0');
      s_space_counter <= (others => '0');
      s_dah_counter <= (others => '0');
      s_dah_done <= '0';
      s_dit_done <= '0';
      s_space_done <= '0';
   if (s_dah_key = '1') then
      s_current_state <= dah;
   elsif (s_dit_key o= '1') then
      s_current_state <= dit;
   else
     s_current_state <= idle;
   end if;
```

The next state we'll talk about is the DAH state. This is the next state in the listing and, as mentioned before, is not necessarily the next state to which the state machine goes. Where the state machine goes will be determined by the conditions for leaving the IDLE state, namely closure of either the DIT or DAH switch. The DAH state code is shown below.

From our state diagram we know that in the DAH state there is a counter involved and that we will remain in the DAH state until this counter has finished counting. The following code segment lists the actions that happen in the DAH state and the conditions under which the DAH state is left, or the conditions under which we remain in the DAH state. There is an output block for the DAH state on the state diagram. When in this state the output of the CPLD is a high level, or a '1'. Notice that the variable 's_out' goes to a '1' in this state. 's_out' is not an output but rather a variable used by the program. In the next-to-last line of the program listing there is an instruction that transfers the information in 's_out' to the actual output variable, called 'output' (declared above in the 'entity' section of the program). This statement is outside of the state machine and is a 'combinatorial' signal. Being a combinatorial signal means that it happens immediately. When signal 's_out' changes, the CPLD output pin, through the 'output' signals, will change immediately. This is how program variables become actions on the CPLD pins. The next line of code determines whether or not the counter has reached the desired count, done by examining the signal 's_dah_done' for a value of '1'.

If the signal 's_dah_done' has a value of '1' this indicates that the counter has reached desired count value and the next instruction causes the DAH state to be immediately left and the SPACE state to be immediately entered. If the signal 's_dah_done' is not a '1' there is no change of state. The next instruction increments the counter. The instructions that follow ask whether or not the counter has reached the desired count and, if reached, sets the signal 's_dah_done' to a '1'. This serves as a 'flag' for the next time the DAH code segment is examined. The next line of code assures that this flag will be examined by telling the program to examine the DAH code segment one more time. If the flag is not set, the 's_dah_done' flag is given a value of '0'. Since none of the conditions for leaving the state have been satisfied, the DAH state code segment will once again be automatically examined (the program remains in the DAH state).

```
when dah =>
        s_out <= '1';
        if s_dah_done = '1' then
          s_current_state <= space;
        else
          s_dah_counter <= s_dah_counter + '1';
         if s_dah_counter = "111111111111" then
                s_dah_done <= '1';
                s_current_state <= dah;
              else s_dah_done <= '0';
```

The same process outlined above for the DAH state also applies to the DIT state. Again, it's not necessarily the next state after idle, but requires the closing of the DIT switch in order to be entered. The code for DIT state is as follows:

```
when dit =>
  s_out <= '1';
        if s_dit_done = '1' then
          s_current_state <= space;
        else
         s_dit_counter <= s_dit_counter + '1';
         if s_dit_counter = "101010101" then
           s_dit_done <= '1';
                s_current_state <= dit;
         else s_dit_done <= '0';
         end if;
        end if;
```

As with the DAH state, there is a counter involved in the DIT state. The program will remain in the DIT state until this counter has reached a count of 341 ('101010101' binary), one third of the value of the DAH state counter. The 's_dit_done' signal will then be set to '1'. On the next occurrence of the clock the 's_dit_done' signal will cause the DIT state to be left and the SPACE state to be entered. The DIT segment, listed above, shows the actions that happen in the DIT state and the conditions under which we remain in the DIT state or leave it. When in the DIT state on the state diagram the output of the CPLD is a high level, or a '1'. Notice that the variable 's_out' is set to a '1' in this state. As with the DAH state, the 's_out' variable is used in a combinatorial statement that determines what the signal named 'output' will pass to the CPLD output pin. Once again, since this is a combinatorial signal, it will occur immediately after 's_out' changes. There are also lines of code that handle the incrementing of the DIT counter and a line of code that asks if the counter has reached the desired count. The variable 's_dit_done' is used to indicate when the DIT counter is done. When the counter has reached the desired count the signal 's_dit_done' receives a '1'. On the next low-to-high transition of the clock, the code for the DIT state is once again examined. This time the 's_dit_done' will contain a '1' and the decision to leave the DIT state and go to the SPACE state will be made.

The next (and last) state to discuss is the SPACE state. This state is entered whenever the 'DIT state' counter or the 'DAH state' counter has finished counting. The signal 's_out' is made to go to zero causing 'output' to go to '0' and the CPLD output pin to go to '0', low level.

```
when space =>
     s_out <= '0';
          if s_space_done = '1' then
            s_current_state <= idle;
          else
            s_space_counter <= s_space_counter + '1';
            if s_space_counter >= "101010101" then
              s_space_done <= '1';
                   s_current_state <= space;
            else s_space_done <= '0';
            end if;
     end if;
   end case;
end if;
```

The code for the SPACE state increments and looks at the SPACE state counter. When the counter matches the desired number of counts (341 counts, in this case) the SPACE 's_space_done' signal is made into a '1'. This is the flag that will be recognized the next time the space state is entered and that will immediately cause the SPACE state to be left and the IDLE state entered. The next instruction causes the state machine to end.

```
end process state_machine;
```

The following instructions are used to synchronize the asynchronous inputs, such as DIT and DAH switch closures, with the CPLD clock. The instructions form two latches ('D' latches) for each asynchronous input. The latches are in series.

```
sync_inputs: process (clk)
  begin
   if (clk'event and clk = '1') then
            s_dit_sync <= dit_in;
                s_dit_key <= s_dit_sync;

                s_dah_sync <= dah_in;
     s_dah_key <= s_dah_sync;
   end if;
 end process sync_inputs;
```

These latches depend upon the clock (a low-to-high transition). The signals 's_dit'sync', 's_dah_sync', 's_dit_key' and 's_dah_key' have all been defined in the architecture section and the inputs 'dit_in' and 'dah_in' have been defined in the entity section of this program. In the attributes, 'dah_in' is associated with CPLD pin 74 and 'dit_in' is associated with CPLD pin 75. On one clock pulse, the 'dit_in' value of '1' is transferred to the signal 's_dit_sync'. Even though there is an instruction immediately after this that transfers the value of 's_dit_sync' to signal 's_dit_key', another clock cycle has to occur before the '1' value of 's_dit_sync' is transferred to 's_dit_key'. Now that it is synchronized, the high level on signal 's_dit_key' is used by the state machine to move from the IDLE state to the DIT state. The same sequence applies for the DAH key. Finally, here is the statement that transfers the state of the 's_out' signal to the 'output' signal. Remember that the 'output' signal is directly related to one of the pins of the CPLD. This happens under the 'attributes' instruction in the 'Entity' section of the program.

```
output <= s_out;
```

The next instruction ends the architecture section of the 'morse_keyer' program.

```
end morse_keyer_arch;
```

This effectively ends the program.

The full MORSE KEYER VHDL code follows:

```
-- File:  C:\Program Files\Cypress\MORSE_KEYER.vhd
--        created: 06/23/08 19:52:13 by R.P.
-- from: 'C:\Program Files\Cypress\MORSE_KEYER'
-- by fsm2hdl - version: 2.0.1.45
--
library IEEE;
use IEEE.std_logic_1164.all;
use IEEE.STD_LOGIC_ARITH.ALL;
use IEEE.STD_LOGIC_UNSIGNED.ALL;

entity morse_keyer is port
      (clk:          in std_logic;
      dah_in:        in std_logic;
      dit_in:        in std_logic;
      output:        out std_logic;
      );

attribute part_name of morse_keyer: entity is "Cy37256P208-83ac";
attribute pin_numbers of morse_keyer: entity is
   "clk:4 output:73 dah_in:74 dit_in:75";

end morse_keyer;

architecture morse_keyer_arch of morse_keyer is
--signal and state machine declarations
  type keyer_states is (idle, dit, dah, space);
  signal s_current_state:  keyer_states;
  signal s_out:            STD_LOGIC;
  signal s_dit_counter:    STD_LOGIC_VECTOR(9 downto 0);
  signal s_space_counter:  STD_LOGIC_VECTOR(9 downto 0);
  signal s_dah_counter:    STD_LOGIC_VECTOR(11 downto 0);
  signal s_dah_key:        STD_LOGIC;
  signal s_dit_key:        STD_LOGIC;
  signal s_dah_done:       STD_LOGIC;
  signal s_dit_sync:       STD_LOGIC;
  signal s_dah_sync:       STD_LOGIC;
  signal s_dit_done:       STD_LOGIC;
  signal s_space_done:     STD_LOGIC;
```

```
  begin

state_machine: process (clk)
begin

  if (clk'event and clk = '1') then

  case s_current_state is
    when idle =>
            s_out <= '0';
            s_dit_counter <= (others => '0');
            s_space_counter <= (others => '0');
            s_dah_counter <= (others => '0');
            s_dah_done <= '0';
            s_dit_done <= '0';
            s_space_done <= '0';
           if (s_dah_key = '1') then
     s_current_state <= dah;
    elsif (s_dit_key = '1') then
     s_current_state <= dit;
    else
     s_current_state <= idle;
           end if;

     when dah =>
             s_out <= '1';
            if s_dah_done = '1' then
              s_current_state <= space;
            else
              s_dah_counter <= s_dah_counter + '1';
             if s_dah_counter = "111111111111" then
                  s_dah_done <= '1';
                  s_current_state <= dah;
                else s_dah_done <= '0';
            end if;
             end if;
            when dit =>
              s_out <= '1';
                 if s_dit_done = '1' then
                   s_current_state <= space;
                 else
                  s_dit_counter <= s_dit_counter + '1';
```

```
                if s_dit_counter = "1111111111" then
                  s_dit_done <= '1';
                      s_current_state <= dit;
                else s_dit_done <= '0';
                end if;
               end if;

      when space =>
           s_out <= '0';
               if s_space_done = '1' then
                 s_current_state <= idle;
               else
                 s_space_counter <= s_space_counter + '1';
                 if s_space_counter >= "1111111111" then
                   s_space_done <= '1';
                       s_current_state <= space;
                 else s_space_done <= '0';
                 end if;
           end if;
          end case;

   end if;
 end process state_machine;

sync_inputs: process (clk)
  begin
    if (clk'event and clk = '1') then
           s_dit_sync <= dit_in;
               s_dit_key <= s_dit_sync;
               s_dah_sync <= dah_in;
        s_dah_key <= s_dah_sync;
    end if;
  end process sync_inputs;
  output <= s_out;
 end morse_keyer_arch;

--That's it!
```

State Diagram and VHDL Coding Conventions

The Following are State Diagram Conventions:

Boxes associated with states are called ACTIONS or ASSERTIONS.

The ENTER ACTION box (>[ ]) indicates one or more actions, or assertions which takes place in each state from which this state is entered. This action is latched and is asserted on the same edge that causes the state to be entered.

The EXIT ACTION box ([ ]>) displays those assertions that must be performed before leaving the present state and going to any other state. This action is latched and is asserted on the same edge that causes the state to be exited.

The TRANSITION ACTION box (T[ ]) displays those assertions that must be performed when a specific guard condition is met. This action is latched and is asserted on the same edge that satisfies the guard condition.

The IN ACTION box ([ ]) is used to define what happens during the state to which it is attached. This is a combinatorial action and is active ONLY while in its associated state. The action is assumed to be defaulted to the opposite logic level in all other states unless explicitly defined by an IN ACTION in one or more of these states.

The GUARDS are those signals (not in boxes) that are located between states. They are the conditions that must be met in order to move from one state to the next.

The following are VHDL Coding Conventions:

If there is no ELSE statement after an IF statement, a Latch is created.

An ELSE statement must follow an IF statement for combinatorial logic to be generated.

All signal lines should be prefixed with an "s_"
Example: s_xxxxx

All constants should be prefixed with a "c_"
Example: c_xxxxx

Low true signals should be followed by "_l" or "_low"
Example: xxxxx_l or xxxxx_low

High true signals are not followed by anything.

Indent two spaces whenever an indent is required.

Signals or inputs that are "low true" should have "_l" (underscore L) placed at the end of their name.

Signals or inputs that are "high true" should have nothing added to the end of their name.

Add comments as to the purpose of a process at the beginning of the process.

Add comments at the end of instructions unless the comment is lengthy in which case, continue the comment on the next line.

The location of all support circuitry (combinatorial or latches) for the state diagram should be referenced on the state diagram.

Before generating a state diagram, state machines should have a detailed description (preferably in words) that outlines the function that the state machine will perform.

Suggestions:
When generating a state machine, make all of the signals HIGH-TRUE. This makes the VHDL code for the state machine more readable. In addition, all of these signals will be automatically initialized to zero.

Synchronizing Asynchronous Signals

This synchronization was done in the VHDL State Machine Example, but it may be helpful to have a schematic representation of synchronizing a signal with a clock.

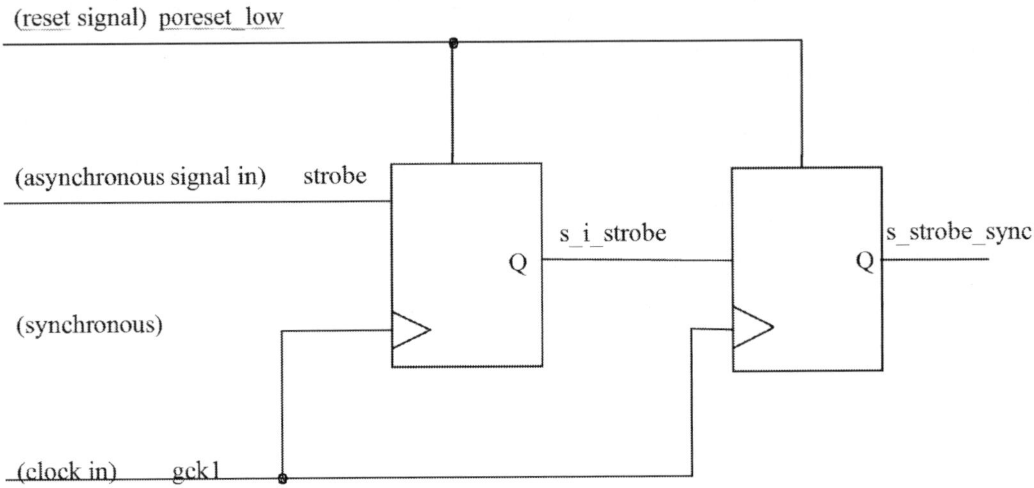


```
--Synchronizing the input 'strobe':

sync_inputs: process (gck1, poreset_low)
  begin
    if poreset_low = '0' then
      s_i_strobe <= '0';
      s_strobe_sync <= '0';
   elsif (gck1'event and gck1 = '1') then
      s_i_strobe <= strobe;
      s_strobe_sync <= s_i_strobe;
   end if;
 end process sync_inputs;
```

The asynchronous input is now synchronized with the clock. That's about it. Get a good book on VHDL programming language and start your application.

Creating a Single Pulse from a Synchronized Signal

The following circuit creates a positive pulse of one clock width, with the falling edge of the s_strobe_sync signal.

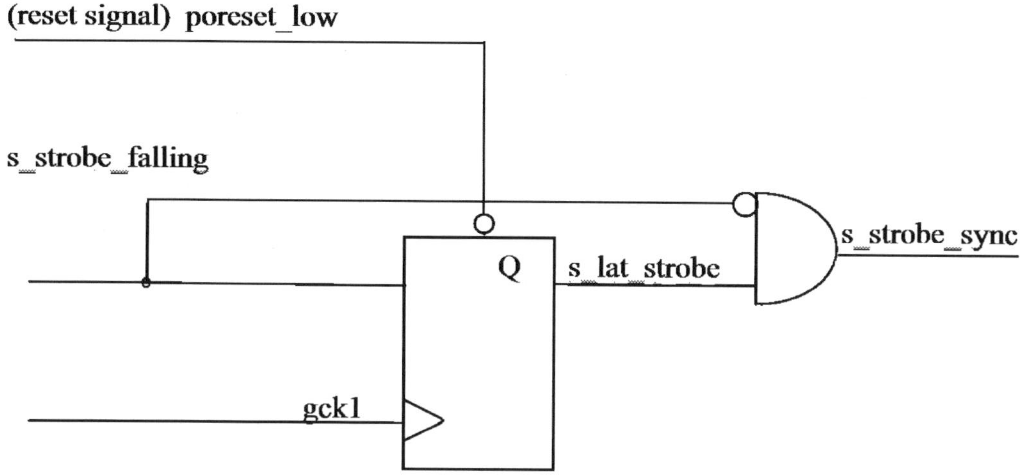

--Handling the input latch for the CPU "strobe" signal (now synchronized in the
--"Synchronizing Inputs" process.

--A low-going pulse on the reset line causes a high (true) out on the s_strobe_sync signal
--line.

```
strobe_latch: process (gck1, poreset_low)
   begin
     if poreset_low = '0' then
       s_lat_strobe <= '0';
    elsif (gck1'event and gck1 = '1') then
       s_lat_strobe <= s_strobe_sync;
    end if;
   end process strobe_latch;

 s_strobe_falling <= not(s_strobe_sync) and s_lat_strobe;
```

NOTE: To create a positive pulse with the rising edge of s_strobe_sync, swap the inverting input of the AND gate to the s_lat_strobe signal.

Hope this helps.

Getting a Degree While You're Working

I had a bit of a leg-up when I went for my B.S.E.E. I had already gone to college for over three years in a Biological Sciences program. Well, when I sat down with the chairman of the Engineering department, it turned out that about a year and a half of courses were not usable toward the engineering curriculum. O.K., I bought that and started classes in the E.E. department. At the time I was married with no kids and had a good job that paid educational benefits.

Things changed along the way. My wife and I bought a house. Then we decided to have a child. My course involvement went from two courses a semester, to one course a semester, to missing every other semester. Then came one great big steaming pile of motivation! Less than two days after my son was born, I got laid off. I scratched for a job and finally got one with a small company that had no educational benefits. Still, I saw the handwriting on the wall and began the two-course-a-semester regimen, paying with my own money. This soon became two to three courses a semester as the push was on. Four years and another job later I finally got the B.S.E.E. Because I had already been working in engineering for so long, I immediately went for the E.I.T. (Engineer in Training) and the P.E. (Professional Engineer) exams, both at the same time. Most of that effort was getting forms filled out and a one summer course toward the E.I.T. No trivial effort, but I had always had a small dream of being a P.E., so why not?

At that point I was in the job/school mode so I applied for Graduate School and started in their M.S.E.E. program. They had several areas of the state in which I could take courses so I doubled-up and tripled up on my courses and wound up completing the M.S.E.E. in less than two years. I'm no genius so this did take a lot away from family time. My wife would take my son out of the house almost every Sunday and I would study the whole day. Evenings were also often long. I wound up completing everything in less than six years. It has definitely paid off.

So what are my thoughts about this? The best time to get a B.S. degree is immediately after high school. You're still in the study mode and you have a lot of company your own age. By the time you go for your M.S. there seems to be more understanding of your situation by the school and faculty. It's an older group of classmates and, in many cases students have full time jobs. This doesn't mean that the work load is any lighter, but there is more latitude for events in your life and, probably because of smaller classes there seems to be more help available.

May the force be with you.

Secrets of the PID Control Loop

The following is intended to share with the engineering community, the calculations necessary to generate an ideal motion profile as well as a brief discussion of PID loop control and some PID loop "tuning tips". The ideal motion profile can also be used in control applications involving fluid flow rates, temperature changes, light intensity variations, plating deposition or erosion and gas pressures changes. Three basic formulas govern the generation of this profile. These formulas lend themselves to being used in any computer or microprocessor based system using any programming language available. The application will define the processor speed required. The use of these formulas and when to apply them is discussed in the following text.

Motion Applications:
Applications involving motion are one of the more common applications of PID controls. Based upon parameters such as velocity, acceleration and move distance, a profile of the ideal move is generated. In closed loop systems (systems using positional feedback information), the motion profile is the result of continuously generating positional information, based upon time, as the time increases from the start of a move to the finish of a move. This positional information results in a profile with smooth acceleration from the start and smooth deceleration to the finish. The positional information of this profile is compared to the actual move position while the move is taking place. Any difference between the profile position and the actual position results in the generation of an error signal. This error signal is acted upon by the PID operators and is used to produce a control signal to the drive unit. The drive unit controls the power to the device which drives the system, such as a motor, solenoid or servo valve.

A short discussion of PID loops:
Several devices exist on the market today which incorporate closed loop system control using an internally generated move profile. These devices range from Programmable Logic Controllers (PLCs) to dedicated Very Large Scale Integrated (VLSI) circuits. The profile is usually used in conjunction with Proportional, Integral and Derivative operators. The values for these "PID" operators are determined by the response of the load. PID values must also remain within the limits of the drive system. The PID parameters operate on the difference between the actual position and the position predicted by the profile for a given time. The result is an error signal which is sent to the drive system and drives the actual position toward the position dictated by the profile. The PID operators are more commonly known as Kp (proportional), Ki (integral) and

Kd (derivative). Most PID controls are digitally based devices. The main advantage of a digital PID control system is accuracy. Encoders and feedback devices with resolutions greater than 1 micron are in common use today. The motion equations which are presented below will allow an ideal profile to be generated through the use of software. In this way, dedicated hardware is not required in a control system. Depending upon the speed of the computer used, several processes or axes of motion can be controlled at the same time.

Figure A (below) outlines a simple control system utilizing a profile generator, feedback signal, PID operators, and Drive Unit.

Figure A

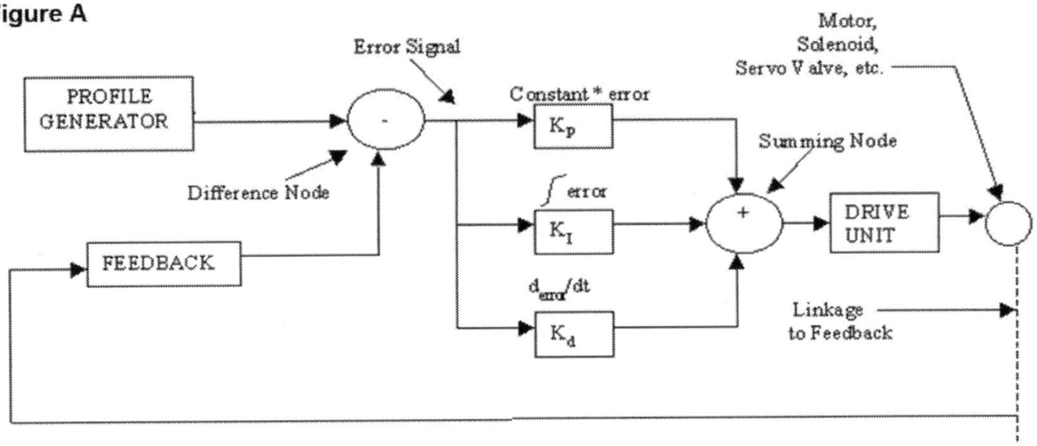

The move path in PID loop control systems is commonly known as a trapezoidal velocity profile or "trapezoidal move" profile. Figure B illustrates a Trapezoidal Move Profile:

Figure B

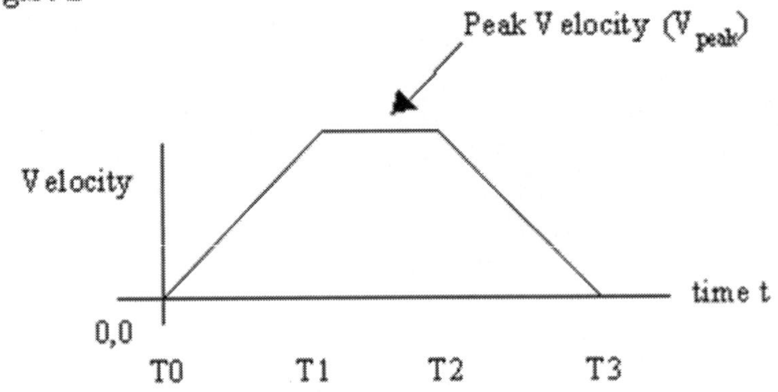

The trapezoidal move figure plots velocity (vertical axis) versus time (horizontal axis). Notice that the velocity is either linearly increasing, linearly decreasing, or remaining unchanged. The trapezoidal move can be broken down into four distinct time periods:

1. T0 The time at which the move profile begins and the velocity begins to increase.

2. T1 The time at which the move profile reaches maximum, or peak velocity and the velocity remains constant.

3. T2 The time at which the move profile begins to decrease in velocity.

4. T3 The time at which the final move position is reached and the velocity goes to zero.

Note: If the maximum or peak velocity is never reached, T1 will occur at the mid-point of the move. The velocity will increase linearly until half of the move has been completed. The velocity will then decrease linearly until the end point of the move has been reached (at T3). In this case, both T1 and T2 will occur at the same time.

Trapezoidal Move that never reaches maximum Velocity (V_{peak}):

Figure C

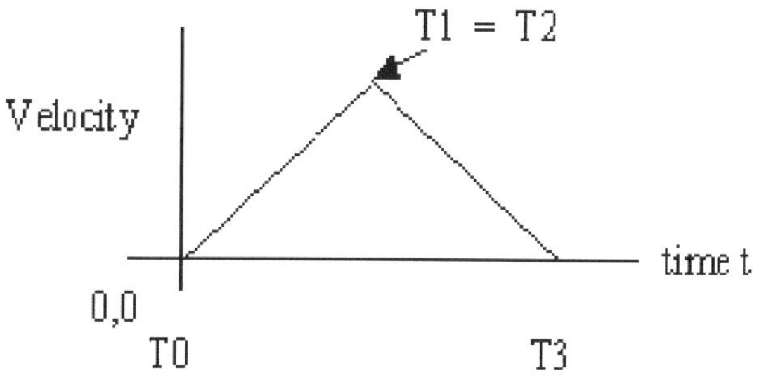

The standard move parameters are Peak Velocity (the maximum Velocity reached during the move), Acceleration, Move Distance and Time required to make the move. The following is an example of how Acceleration and Move Distance between times T0 and T1, can be determined given the Peak Velocity and the Time required to make the move:

Given the peak velocity (V_{peak}) and the time period from T0 to T1 determine the distance traveled, the average velocity and the acceleration of a move.

Figure D

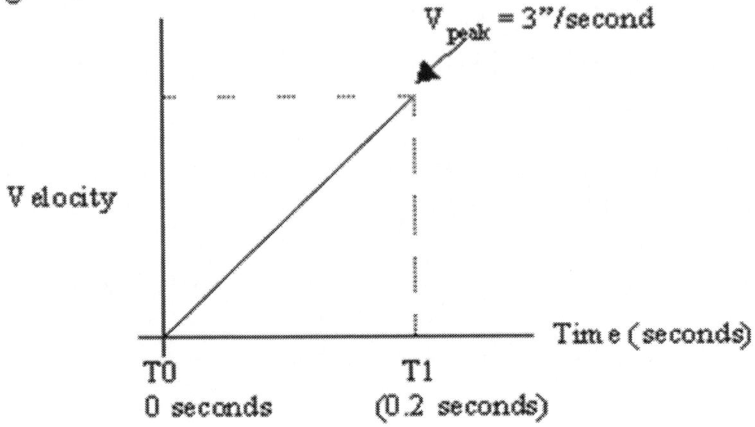

Notice that V_{peak} is reached at T1. Because the velocity is linear, the average velocity ($V_{average}$) is half the peak velocity (V_{peak}):

1) 3"/second divided by 2 = 1.5"/second = $V_{average}$

The distance traveled between T0 and T1 is calculated as follows;

2) Distance traveled, T0 to T1 = $V_{average}$ * (T1 - T0), which is velocity * time

Letting T0 equal 0 (the start time of the move):

= $V_{average}$ * T1

= 1.5"/sec * 0.2 seconds

= 0.3"

Figure E

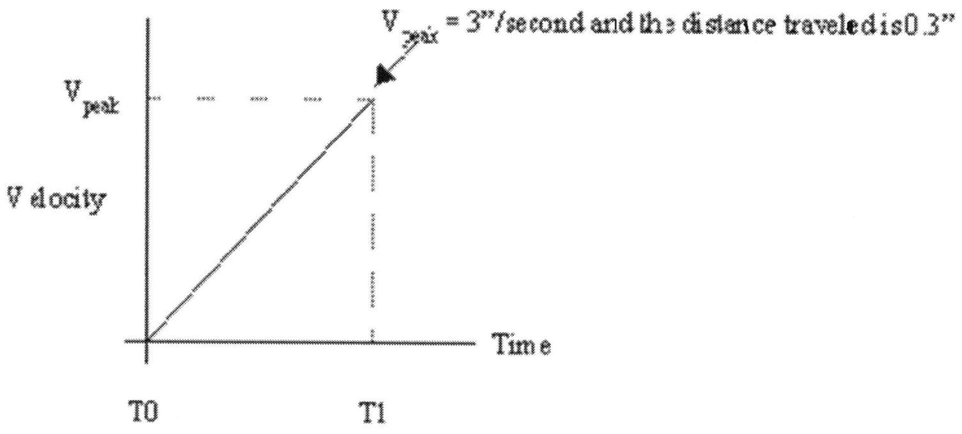

If the distance traveled from T0 to T1 had been given, here is another way of calculating the average velocity:

(3) Average velocity = ds/dt

= 0.3"/0.2 seconds

= 1.5"/second

In either case, the acceleration can be calculated as follows:

(4) Acceleration $= dV_{peak}/dt$

$= \dfrac{3\text{"/sec}}{0.2 \text{ seconds}}$

= 15"/sec/sec or 15"/sec^2

MOVE PARAMETERS:

The most common parameters given for any move are:

1. Peak Velocity
2. Acceleration
3. Total move distance

Using these three parameters, the position at any time "t" for a trapezoidal move can be calculated:

Trapezoidal Move Profile:

Figure F

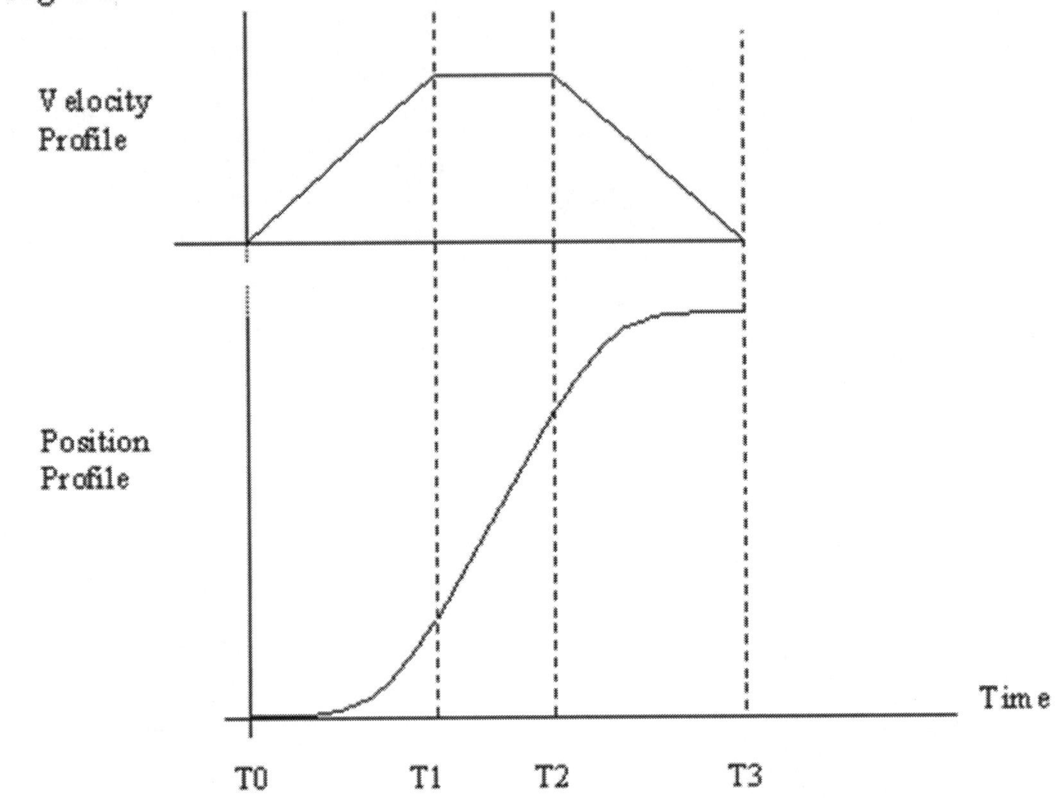

Calculations given Peak Velocity, Acceleration and Move Distance

As a check, the following calculations will use the previously determined Peak Velocity (3"/second) and Acceleration (15" /sec^2). The initial distance will be zero, as will the initial time, T0. Assume a total move distance of 0.8".
T1 is calculated as follows:

(5) Using the Physics formula[1]: $\text{Velocity}_{peak} = \text{Acceleration x Time})$ and letting 'Velocity_{peak}' be represented by the symbol 'V_{peak}':

$$T1 = V_{peak} / \text{Acceleration}$$

$$= \frac{3''/\text{sec}}{15'' / \text{sec}^2}$$

$= 0.2$ seconds (This checks with the value given in Figure D)

(6) Also from Physics[1]: " Distance = ½ * Acceleration * time^2 "

Using this formula, the distance from T0 to T1 can be calculated:

$$\text{Distance traveled, T0 to T1} = \tfrac{1}{2} * (15'' / \text{sec}^2) * (\text{T1})^2$$

$$= (7.5'' / \text{sec}^2) * 0.04 \text{ sec}^2$$
$$= 0.3''$$

(this result checks with our initial determination of the distance from T0 to T1 as defined in equation (2)).

This same formula can be used to calculate the distance at any time "t" between T0 and T1. The formula below is the general form of the distance equation for any time "t" between T0 to T1 for any move.

(7) Distance traveled at any time "t" between T0 and T1= ½ * Acceleration * t^2

At this point, a decision must be made. If the "Distance at T1" multiplied by two is greater than the Total Move Distance, this indicates that the maximum Velocity is never reached during the move. The distance at T1 automatically becomes the Total Move Distance divided by two, and T1 equals T2.

When: 2 * Distance at T1 > Total Move Distance

(8) Then: Distance at T1(and at T2, since T1 equals T2) $= \frac{\text{Total Move Distance}}{2}$

For those cases where the distance traveled from T0 to T1 is <u>less than or equal to</u> the Total Move Distance divided by two, the distance traveled from T0 to T2 is calculated using formula (9). In our example this happens to be the case, that is;

(9) Distance traveled (T0-T1) = Total Move Distance - Distance traveled T0 to T1

$$= 0.8'' - 0.3''$$
$$= 0.5''$$

This result is greater than zero, therefore this move does reach maximum velocity (T1 and T2 are not the same).

For those cases in which the distance traveled from T0 to T1 is greater than

$$\frac{\text{Total Move Distance}}{2}$$

This indicates that the maximum velocity (V_{peak}) is never reached. For moves where V_{peak} is not reached, T1 = T2. Therefore T3 = 2 * T1, = 2 * T2. This also means that the distance, T1 to T2 = 0.

The calculation for T1 when maximum velocity (V_{peak}) is not reached, is as follows:

Re-writing formula (6) yields;

$$T1^2 = \frac{\text{Distance traveled from T0 to T1}}{\frac{1}{2} * \text{Acceleration}}$$

$$T1^2 = \frac{\frac{\text{Total Move Distance}}{2}}{\frac{\text{Acceleration}}{2}}$$

(10) $$T1 = T2 = \sqrt{\frac{\text{Total Move Distance}}{\text{Acceleration}}}$$

(when Vpeak is never reached)

Calculating time T2 :

(11) $$T2 = \frac{\text{Distance traveled, T0 to T2}}{V_{peak}} + \frac{T1}{2}$$

Substituting the given values (above):

$$= \frac{0.5"}{3"/\text{second}} + \frac{0.2 \text{ seconds}}{2}$$

$$= 0.26666 \text{ seconds}$$

Calculating T3, the Total Move Time, for any move:

(12) T3 = T2 + T1 (also true when T1 = T2, like moves that don't reach V_{peak})

= 0.26666 + 0.2

= 0.46666 seconds

Calculating the Distance traveled, T1 to T2, (when T1 $\neq$ T2):

(13) Distance traveled (T1→T2) =

Distance traveled (T0→T2) - Distance traveled (T0→T1)

= 0.5" - 0.3"

= 0.2"

Formula (14), below, is the general formula for calculating Distance traveled at any time "t" between T1 and T2. This formula is linear. It follows the slope, at time T1, of the curve which generated the distances between T0 and T1 plus the offset of the distance at time T1.

(14) Dist. traveled (T1→T2) = V_{peak}* ($t_{present}$ →T1) + Dist. traveled (T0→T1)

Calculating the Distance traveled, T2 to T3:

(15) Distance traveled (T2→T3) = Distance traveled (T0→T1)

= 0.3"

Formula (16), below, is the general formula for calculating Distance traveled at any time "t" between T2 and T3. The second term of this equation is an exponential function which decreases toward zero as $T_{present}$ approaches T3. This function is subtracted from the Total Move Distance. Notice that at the end of the move, $t_{present}$ equals T3 and nothing is subtracted from the Total Move Distance.

(16) Distance, at any time "t", between T2 and T3 =

$$\text{Total Move Distance} - \frac{\text{Acceleration} * (T1 - t_{present} + T2)^2}{2}$$

Note: When calculating a move profile using equation (16), don't forget to terminate the calculation at the final move position. If further calculations are made, the move profile will begin to decrease back toward the original position.

Summary of the results of the example:
The given parameters for this example were:

Total Move Distance	=	0.8"
Velocity (max)	=	3"sec
Acceleration	=	15" /sec^2

The calculations resulted in the following:

Distance from T0 to T1 = 0.3"
Distance from T1 to T2 = 0.2"
Distance from T2 to T3 = Distance from T0 to T1 = 0.3"
Distance at T3 (Total Move Distance) = 0.8"
T1 = 0.2 seconds
T2 = 0.26666 seconds
T3 = 0.46666 seconds

Two additional examples:

An example when maximum velocity is never reached;

Given:

Total Move Distance	=	0.9"
Velocity (max)	=	15"/sec
Acceleration	=	230"/sec^2

Find: T1, T2 and T3

$$T1 = \frac{\text{Velocity (max)}}{\text{Acceleration}} = 0.0652 \text{ seconds}$$

$$\text{Distance from T0 to T1} = \frac{\text{Acceleration} * T1^2}{2} = 230 * (0.0652)^2 = 0.489"$$

Determining whether or not Velocity (max) is reached:

Distance (T0 to T1) * 2 = 0.978"

Since Total Move Distance is less than 0.978", the move never reaches Velocity (max).

$$\text{Therefore: } T1 = T2 = \sqrt{\frac{\text{Total Move Distance}}{\text{Acceleration}}} = 0.0625 \text{ seconds}$$

$$\text{The Distance (T0-T1)} = \text{Distance (T0-T2)} = \frac{\text{Total Move Distance}}{2} = 0.45"$$

T3 = T1 + T2 = 0.125 seconds

Here's an example when maximum velocity is reached;

Given:

Total Move Distance = 1"

Velocity (max) = 15"/sec

Acceleration = 230"/sec^2

Find: T1, T2 and T3

$$T1 = \frac{\text{Velocity (max)}}{\text{Acceleration}} = \frac{15}{230} = 0.0652 \text{ seconds}$$

$$\text{Distance (T0 to T1)} = \frac{\text{Acceleration} * T1^2}{2} = \frac{230 * (0.0652)^2}{2} = 0.489$$

Determining whether or not Velocity (max) is reached:

Distance (T0 to T1) * 2 = 0.978"

Since this is less than the Total Move Distance, the move does reach

Velocity (max).

Therefore, the results of the formula " $T1 = \frac{\text{Velocity (max)}}{\text{Acceleration}}$ " will be used.

$$\text{Distance (T0 to T2)} = \text{Total Move Distance} - \text{Distance (T0 to T1)}$$

$$= 1 - 0.489, \quad = 0.511$$

$$T2 = \frac{\text{Distance (T0 to T2)}}{\text{Velocity (max)}} + \frac{T1}{2} = 0.067 \text{ seconds}$$

$$T3 = T1 + T2, = 0.132 \text{ seconds}$$

Summary of the Move Profile Equations:
To find distance, with respect to time, for all times "t" between T0 and T1, use the formula:

Distance when time "t" is between T0 to T1 $= \frac{1}{2} * \textit{Acceleration} * t^2_{present}$

To find distance, with respect to time, for all times "t" between T1 and T2 (providing V_{peak} is reached during the move), use the formula:

Distance when time "t" is between T1 to T2 $= V_{peak} * (t_{present} - T1) +$ ***Distance traveled, T0 to T1***

Note: When V_{peak} is never reached, the distance, with respect to time, for all times "t" between T1 and T2 is equal to '0'.

To find distance, with respect to time, for all times "t" between T2 and T3, use the formula:

Distance when time "t" is between T2 to T3 =

$$\textit{Total Move Distance} - \frac{\textit{Acceleration} * (T1 - t_{present} + T2)^2}{2}$$

Determining PID parameter settings in motion control systems:

An error signal is generated when the positional feedback (from the unit or system being controlled) differs from the calculated move profile position for any time, "t" between T0 and T3. This error is multiplied (Proportional) by some number Kp, Integrated (Integral) over some time period Ki , and checked against previous errors to get the rate of change (Derivative) of error Kd. All of these PID factors are summed together to generate a control signal which tries to reduce this error. This control signal is sent to the Drive Unit. The values of Kp, Ki and Kd are based upon the system parameters such as the inertia of the load, the mechanical response of the system and/or the electrical response of the system. In most applications there are limits to the ability to calculate Kp, Ki and Kd based upon the system response because of the vast number of variables involved. For this reason, Kp, Ki and Kd are usually set empirically. A trial-and-error method results in settings which "work" but may show instabilities under certain move conditions. It is worth spending the time to become familiar with system behavior when optimizing PID control loop parameters. Several programs used in control systems are capable of "learning" the parameters of the load by exciting the system with a move resembling an impulse function and then observing the response of the system. The PID parameters are then automatically varied in order to match the load and move parameters.

What to do when a move parameter changes during a move.

In systems controlled by computers, move profile calculations can be made "on the fly". **The only parameter that should change while a move is in-progress is the end-point of the move.* The only parameter affected is the Total Move Distance. Changes in peak velocity or acceleration during a move can be disastrous because of the effect on the dynamics of the load and drive system. Changes in these parameters cause the control parameters Kp, Ki and Kd, which were determined for the original move, to no longer apply. Here are some simple rules regarding "on the fly" changes in move distance:

1) If the end-point of the move is increased while the move is between T0 and T1, it is possible to calculate a new T1, T2 and T3 and generate a new move profile.

2) If the end-point of the move is decreased while the move is between T0 and T1, it must first be determined if the new end-point is equal to, or greater than twice the present position. If it is equal to, or greater than twice the present position, the move can be re-calculated. If it is less than twice the present position, the action is to define the present position as being at T1 (because V_{peak} was not yet reached, T1 = T2) and recalculate the position for the remaining half (T2 to T3) of this shortened move. Once the end-point of this shortened move has been reached, calculate an entirely new move to the new end-point.

3) If the end-point of the move is increased while the move is between T1 and T2 (indicating that Vpeak has been reached), the move profile can be re-calculated and a new T2 and T3 determined.

4) If the end-point of the move is decreased while the move is between T1 and T2, the move should immediately be considered as having reached T2 (shown as T2' in Figure G) and a determination as to whether or not the time from T2 to 'new' T3 (move time to the new endpoint), is the same or greater in length than the period T0 to T1 (new T3 $\geq$ 2 * new T2). If it is greater, the move profile can be re-calculated and a new T2 and T3 determined. If it is less, then the present move profile should be re-calculated with $t_{present}$ equal to T2. The move should then be completed. This will finish the present move as quickly as possible. Once the end-point of this shortened move has been reached, calculate an entirely new move to the new end-point. Refer to 'Figure G', below for an illustration of this situation.

Figure G

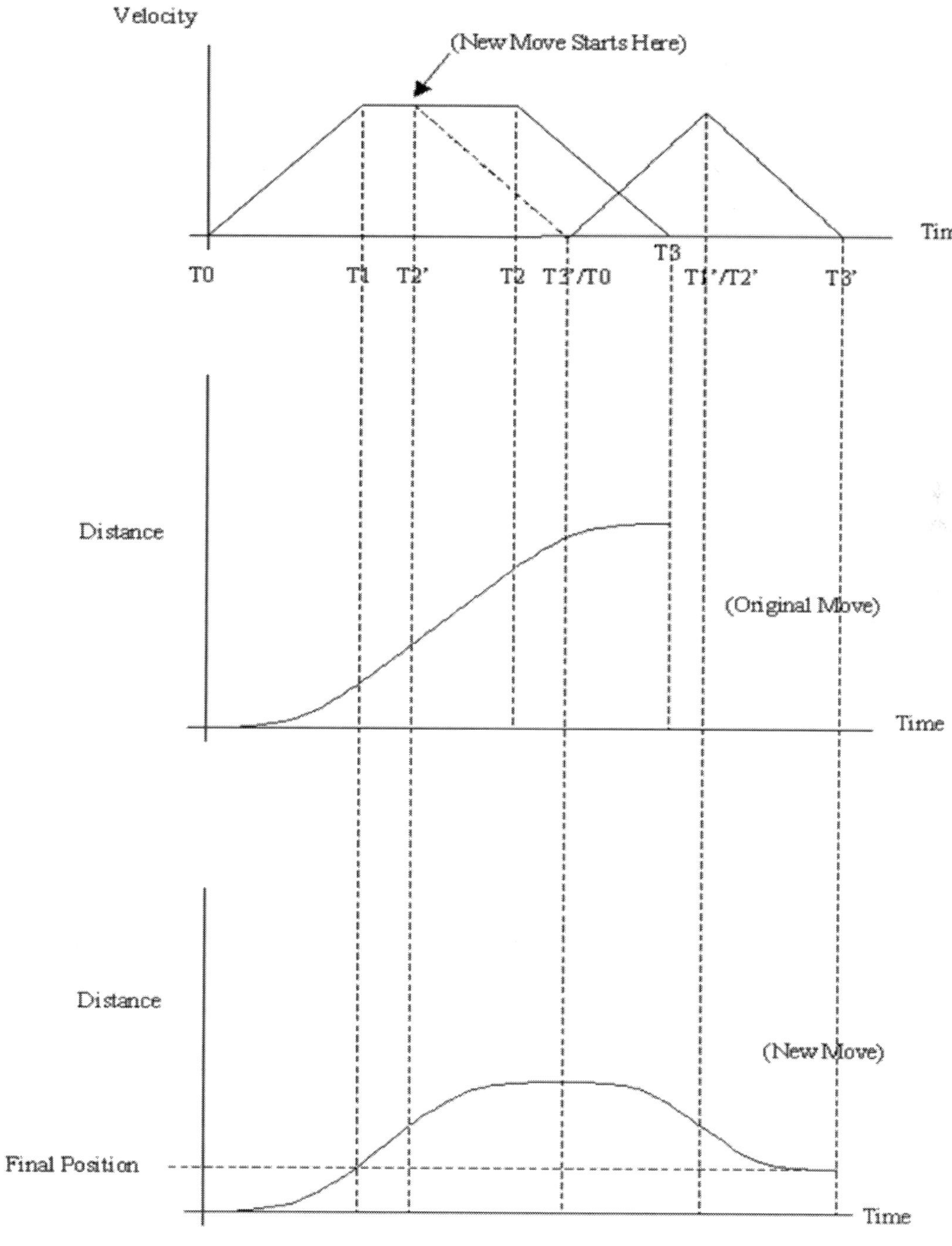
Velocity
(New Move Starts Here)
Time
T0
T1
T2'
T2
T3'/T0
T3
T1'/T2'
T3'
Distance
(Original Move)
Time
Distance
(New Move)
Final Position
Time

Notice that the velocity profile is always positive and is therefore independent of position or direction of move. The "Position" graph of 'Figure G' shows a reversal of direction while the "Position" graph of 'Figure I', below shows a continuation in the same direction.

5) If the end-point of the move is increased while the move is between T2 and T3 the move profile can be re-calculated. Refer to 'Figure I', below:

Figure I

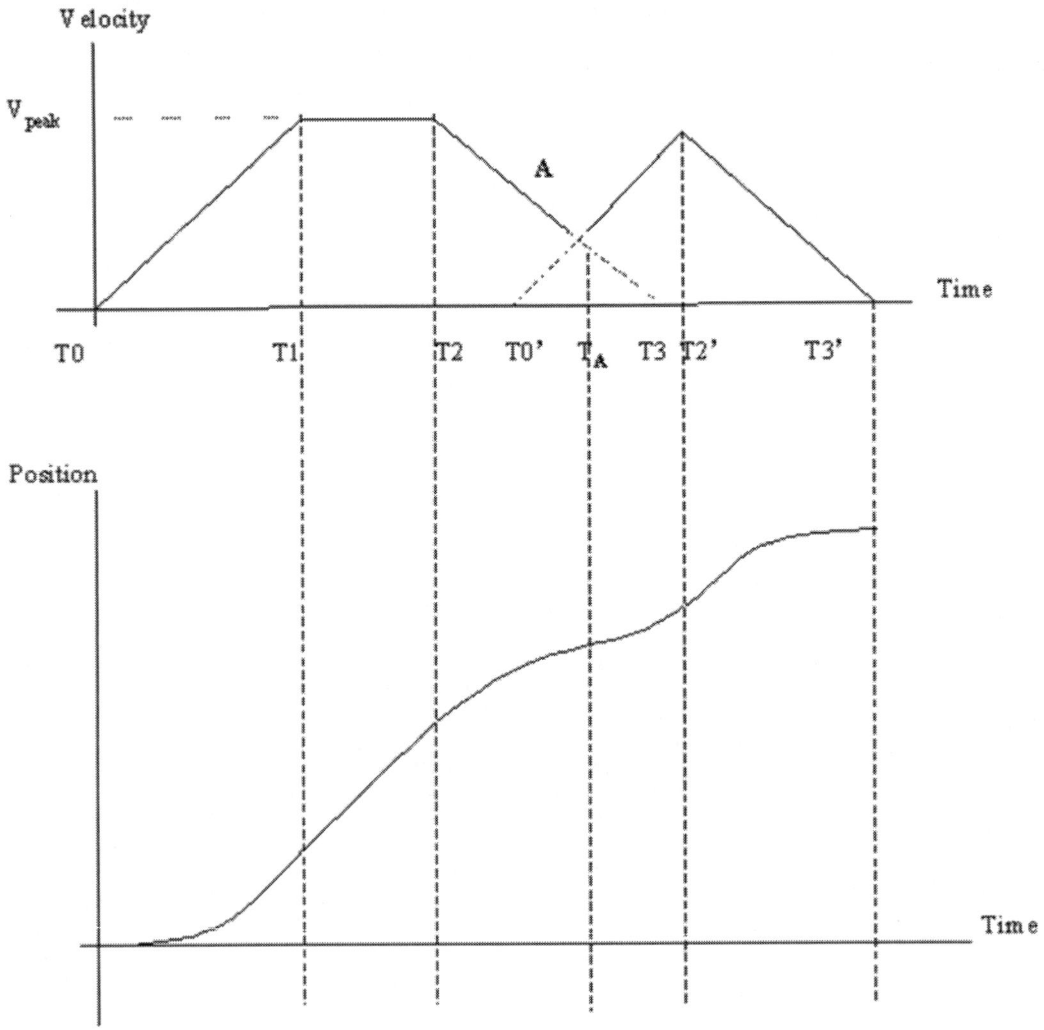

This actually involves calculating a new move profile while the initial move profile is in progress. Point **A** is the point of intersection of the new move profile with the old move profile. The remaining distance between **A** and T3 will be the same distance as from T0' to **A**. Therefore;

*2 * (distance, time T_A to time T3) + (distance at T3' - distance at T3) = Total Move Distance (new move)*

Time T0' has already occurred. Using the Total Move Distance for this continuing move, calculate the new velocity changes (T0', T2', T3'). Begin the new move at the position calculated for T_A (which will be the present position). Refer to Figure 'I' below:

6) <u>If the end-point of the move is decreased while the move is between T2 and T3</u>, complete the move profile, then calculate a new move profile using the present position, the position at T3 (end of first move) and the new end-point.

Summary and Tips about PID tuning:

One way of determining whether or not a PID control loop is out of control, is to observe the input signal to the Drive Unit. If this signal *ever* exceeds the input signal range, then the control system does not have enough drive to control the load with the present parameters (Kp, Ki and Kd, Velocity and Acceleration). Reducing acceleration is usually the first thing to try.

Adjusting PIDS:

Present drive systems usually have an 'auto-tune' feature that is supposed to automatically tune the system for the user. In many cases, auto-tuning is adequate but in those cases where you have to make a lot of time consuming keyboard entries in order to tune PIDs, here are some tips on minimizing set-up time:

a) Remember to observe the input to the Drive Unit. This signal should *never* go out of range. The usual control signal range is +/- 10 Volts. Observe this input with a high bandwidth device, such as an oscilloscope and make sure it remains in-range.
b) Only adjust one parameter (Kp, Ki, Kd, Velocity or Acceleration) at a time.
c) Keep track of the settings and system response to these settings as adjustments are made.
d) Zero out Kp, Ki and Kd.
e) If the time to settle to final position is important, enter a very small value for Ki, the "integral" parameter. The primary use of Ki comes about after the completion of the move. At that time, Ki will help to keep the system at final position. However, during a move, Ki acts as an error storage parameter and can adversely affect settling to the final position. Using Kp, Kd and only a small amount of Ki will get you close to the proper system response and will greatly help final position settling time.
f) Increase Kp until system drag is overcome.
g) Increase Kd until the system response is very tight (sometimes evidenced by an audible hum or squeal) and then slightly reduce Kd.
h) Continue to adjust Kp and Kd until the move is as smooth as possible and the time required for settling to the final position is minimal.
i) Introduce a small amount of Ki, but observe the effect on settling time. If positional accuracy is not important (the system can settle to a position plus or minus a fairly wide range) it may be possible to increase Ki and thus reduce errors in final position.

A note about settling time:

The ideal situation at the end of a move is to reach the end point of the move with both the velocity and the error signal being zero. This is seldom the case. In systems where a high degree of accuracy is required the end position may have a tolerance of only a few microns. The parameter that contributes the most to long settling times is Ki. This is because Ki is the summation of past errors and this memory has to be forgotten, or cleared before the error signal can go to zero. For short moves, or moves where peak velocity is never reached, the time to settle to a high tolerance final position could approach 1/3rd the time of the entire move. When settled at the final position, Ki can be extremely useful in remaining at this final position. Any small error will be integrated and will eventually return the system to within the limits of the final position.

One parameter not discussed above is the "Velocity Feed Forward." This parameter is simply a constant multiplier of the error signal that is only in effect when there is a change in velocity. For a trapezoidal move, Velocity Feed Forward is in effect when the velocity increases (T0 to T1) and when the velocity decreases (T2 to T3). The added gain, as the velocity decreases to the end of the move greatly helps to reduce the settling time.

[1.] Fundamentals of Physics, Halliday, David; Resnick, Robert; Walker, Jearl
July 1, 1999, Published by John Wiley & Sons ISBN 0471253820

Happy controlling!

The RIGHT Circuit

So there you are. Someone has given you the task of creating a widget. Maybe you get the specs, maybe you have to make the specs. It doesn't matter what it does; control something, display something, whatever. You start noodling and doodling and after a couple of hours or a couple of days you decide on the best way to approach the problem, i.e. what discipline to use. Here's where it sometimes gets hairy. You can start with the basic building blocks of the method you have chosen or you can try to be a bit original. Believe me, solve it with the building blocks first. You can be as original as you want to be after you have something that works. About the most common constraint given to the designer is 'cost'. The darned thing has to cost under some fixed amount. So even after you have a working circuit you may be far from done. You may even have to go back and select a different approach (digital instead of analog or vice-versa).

The next big physical restrictions are 'size' and 'weight'. Unless you're the one designing the packaging, you have to start interfacing with the person who is packaging this widget. They have their own set of problems (COST, materials, mounting, structural integrity…COST). The last, but probably the hardest constraint is Time to Market. They forget that you have to design it, build it, and make it work every time, all the time.

O.K., you finally have a widget you think fits the bill. Next step is Testing! Don't leave this step out or the people using this widget, the guys in the field will do it for you and heaven help you if the guys in the field turn out to be the customer. Customers know junk when they see it. To them it's anything that doesn't work every time, all the time. So find out the environments in which it will be used and test the heck out of it.

I hope you have been documenting this 'widget' development process as you have been going along. The whole idea is to be able to go back and see why you did things a certain way. Why? Because if this widget is successful you're going to make a lot of them for a long time and I haven't seen a widget yet that didn't have some weak point in the circuit design show up weeks, months and years later. Don't forget to put time for this activity in the bottom-line cost of this widget.

So if you've done a good job AND you're lucky, the widget will be a success and you can start letting the customers drive the next generation. When your customer says "It's great but I need it to do this too". Then you know you're on the way.

Relay Ladder Logic

Electrical Engineers. Hey! What's up with that? Most of the engineers I know have expertise in electronics. Analog, Digital or Software but not electrical. Electrical involves high voltages, lots of current, motors and generators. Not so 'weed hopper!'. This whole electronics thing started with electrical guys (and some gals). They were using relays as logic elements for years and years before the integrated circuit came along. By the way, they still are. Most of the production done in the world depends upon the logic of relays. How about a few basics.

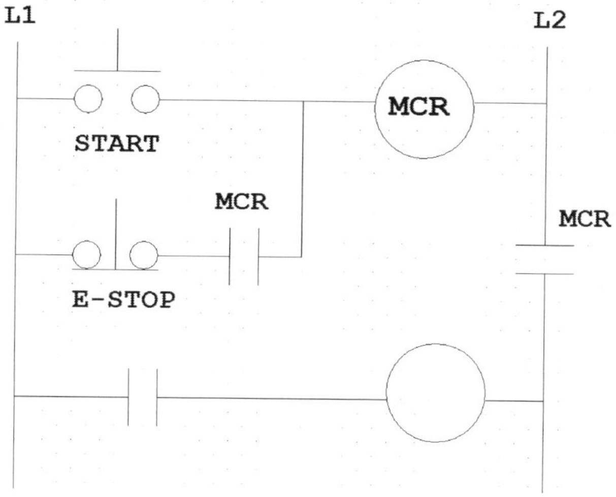

Relay logic involves a two rail system, one rail is called L1 and the other is called L2. Electronics guys would normally refer to this as Power and Ground. Everything happens between L1 and L2. The voltage between L1 and L2 can be just about anything. The most common voltages are +28VDC (also called "Control Voltage") and 110VAC. On a relay logic diagram the coil of the relay is indicated by a circle with the name of the relay inside. Contacts that are open when the relay is not energized are known as 'Normally Open' contacts. Contacts that are closed are known as 'Normally Closed'. Normally open contacts are diagrammatically indicated by two parallel, vertical bars and normally closed contacts are indicated by two parallel, vertical bars connected by a slash.

At the beginning of any relay logic scheme is a way of turning on the logic and some way of turning off, or resetting the logic. <u>The Master Control Relay</u> or MCR is what is normally controlled. There is a push button used to pull in the

MCR (START) and one, or more to drop out the MCR (E-STOP). The contacts of the MCR itself control power to the rest of the circuitry in the ladder.

Relay ladder logic formed the basis for Programmable Logic Controllers, or PLCs for a long time. PLCs, although micro-processor based, still incorporate ladder logic but have leaned more toward programming language lately. Now between L1 and L2 are all kinds of decision making stuff. There are; drum timers, multi-byte comparators, lines and lines of 'C' code and much more. Enough of that, let's look at some ladder logic.

How about a simple problem. Two buttons, one of which must be depressed and released before the second button is pushed in order to get an output. The output will be the 'dry contact' (unpowered contact) of a relay. Sounds simple. Believe me, nothing is simple.
The first thing to do is to put in the MCR circuitry. Then add the push-button switches, PB-1 and PB-2.

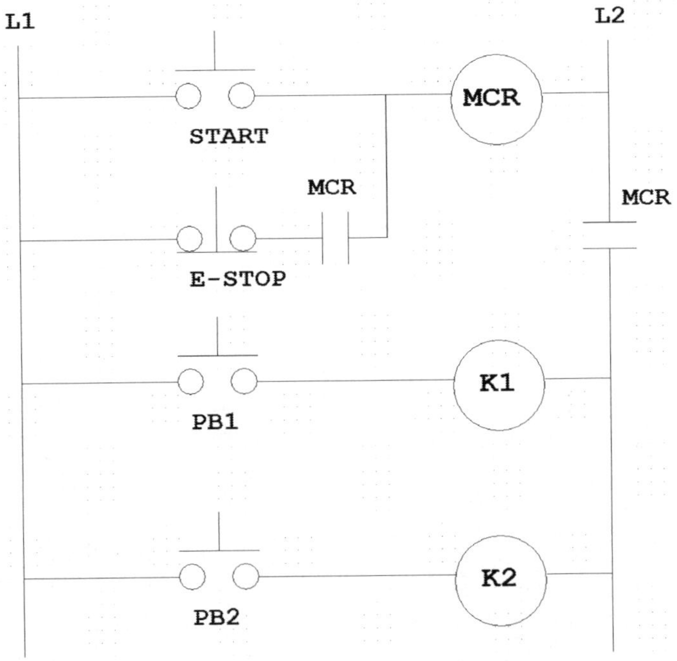

In order to tell that a push-button has been depressed the best thing to do is to latch the signal. So let's latch PB-1 and PB-2.

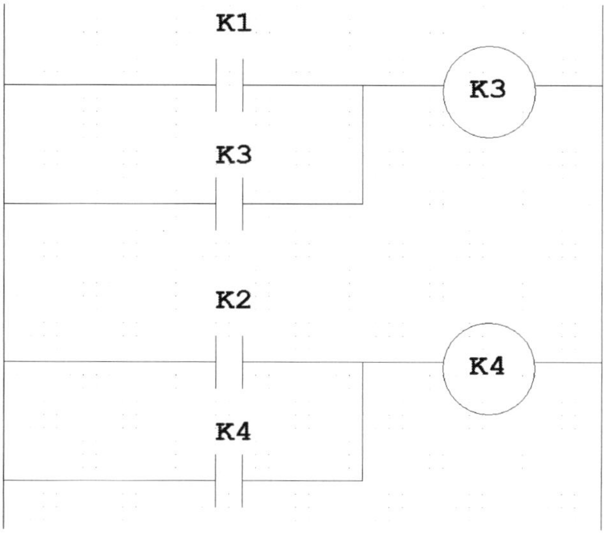

After this, the next step is to make some logic to indicate when PB-1 has been depressed and then released and, at the same time only function when PB-2 hasn't been depressed yet.

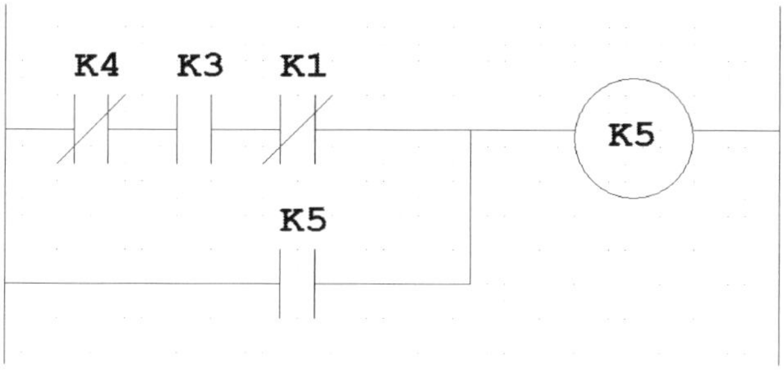

So let's see! When PB-1 is depressed it pulls in K1 which is in turn latched by K3. So PB1 must have been depressed in order for K3 to pull in and complete its portion of the current path that will pull in K5. How about K4. K4 is the latch for K2 which in turn is pulled in as soon as PB-2 is depressed. So in our current path, PB2 must not have been depressed or the normally closed contact of K4 would be open. Now for K1. K1 is directly dependent upon PB-1. If PB-1 is open, K1 is open so if PB-1 is not depressed the normally closed K1 contact will complete the circuit to pull in K5. K5 will then latch itself.

After determining that PB-1 has been depressed and released before PB-2 has been depressed we still have to depress PB-2. K6 will do this for us and then latch itself.

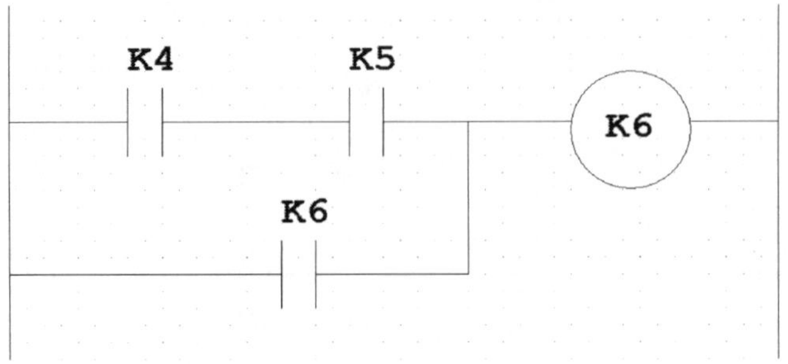

There are other conditions that are not handled in this circuit such as; if PB-2 is depressed before PB-1 then K5 will never pull in. Also, there is no circuitry for determining if PB-1 is depressed multiple times before PB-2, although this is not a pre-requisite for the given problem. However, if PB-2 has been depressed first, the circuit can always be reset by depressing the MCR 'E-STOP' push-button.

The whole circuit looks like this:

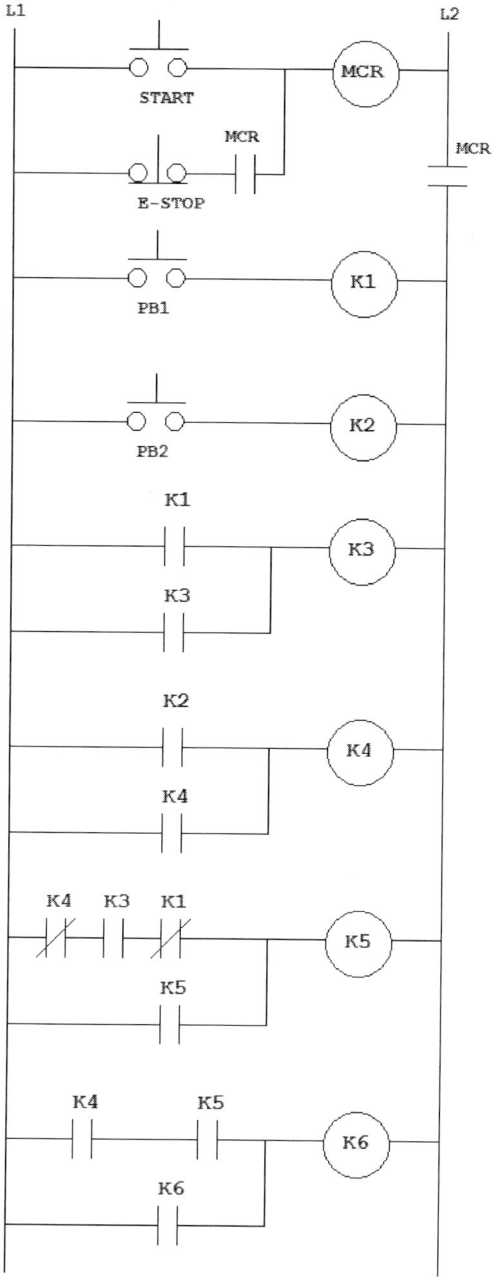

You may want to figure out the missing logic on your own.

As usual, good luck.

The Microprocessor Story

Ok, Ok! This story is about me. It all started back when Houston, Texas was a hot-bed of technology, the up-and-coming 'place to be'. I moved there for a job. Times were so good that I had three job offers in one week. That has never happened since, and probably will never happen again. Anyway, I took the job I though would be the most interesting. As far as I know, I was right. My background at the time was at technician level. I had received military training in electronics but no college training in engineering. Still, I got a position as an engineer. The company I worked for was in the geophysical industry. Like most of the industry in Houston at that time, they searched for oil. Because I had a radar and video background, they plugged me into a video project that gathered geophysical data on a reel-to-reel video tape machine. Quite interesting. But enough of that. I also had computer main-frame experience and it wasn't long before I was approached to take on a project involving a four (4) bit micro-processor with associated hardware. I wondered how a four bit processor worked since the mainframes I had worked with used 32 bit words and 64 bit double words. What the heck could you do with four bits? Well, it turned out that you could do a lot. It was actually an 8 bit machine that multiplexed 4 bits at a time. No problem. Now we had the capability for at least 256 instructions. I think it had about 32 instructions. Sounds something like an early RISC processor doesn't it?

It turned out that these processors and support chips were originally intended for use in bank ATMs but that they had not worked out. So I had this single-board hardware platform to work with. Problem was, I needed to program the EPROMS in machine language, instruction by instruction. The development system allowed me to enter hexadecimal information and to program EPROMs, but that was it. There was no assembler, nothing. Anyway, I wrote a program that was 800 bytes long not including 200 bytes of look-up table. The application was a security system. The I/O consisted of a real-time clock, 128 switch inputs, a card reader input, a key-pad input, an 8 digit 7-segment display and an alarm out. The date and time of the recognition of any input could be programmed from either the keypad or card reader. It was a great little project and the customer was happy.

This was a fun project. Much of the fun was cramming the whole thing into 1024 bytes. This is something that most programmers don't have to worry about today. Program memory has become the least of the problems. Anyway, shortly thereafter the 6800 series of microprocessors came out. I bought an evaluation board with a 6800 and some family support devices on it. The thing was designed for a teletype interface but I didn't have a teletype at the time. I decided to reprogram the board for a hexadecimal keypad with 'address', 'store' and 'increment' keys included. I read the literature and took a stab at programming the keypad through a Programmable Interface Adapter (PIA) chip. I had to use that Intellec-8 system to program an EPROM and had to change the hardware to take this type of EPROM. It took a couple of tries, but I finally got the board to recognize the keypad. After that, I was on the way. I could create programs and check out small routines with ease.

Next I had to create some way to save the information. Using the EPROM specifications, I made an EPROM programmer board and added it to the evaluation board. Great! Now I could create programs, save them and then program an EPROM for use with another 6800-based system, or any processor out there for that matter, all with a hex keypad, an 'Address' entry key, a 'Store' key and an 'Increment' key. I created several dedicated micro-processors with this set-up but finally opted for a teletype entry. Great I/O! It had data entry, paper-tape for saving and re-loading information and a hard copy printout. O.K., O.K! Sounds stupid compared to today's P.C.s and palm-tops, but at one time, this was state-of-the-art, kind of like the wood stove.

All of this gave way to the P.C., especially the open architecture of the IBM P.C. They gave away the farm with that move, but it sure started the computer industry for the common man. Hats off to IBM.

Life is good.

Calculations for a Bridge Circuit

Bridge circuits are usually used in measurement systems in which sensors cause imbalances in known voltage ratios. The advantage of using a bridge is that any changes that are external to the circuit can be detected, that is; if one of the wires of the bridge becomes open circuited or short circuited, a component changes value or if the sensing device is open or shorted. All of these will cause imbalances in the bridge circuit. Voltage drops that are out of the range of the sensor will then be detected.

The calculations for determining the voltage across a sensor within a bridge are in the description that follows. The method of calculation used is called the Thevenin Equivalent method. This involves looking at the circuit in different ways. Calculations are made with the load resistance opened (removed) in order to determine the Thevenin voltage across the load resistance, and then with the voltage source shorted and the resistance of the circuit calculated when LOOKING INTO the load resistance terminals.

Here's an example of a bridge circuit.

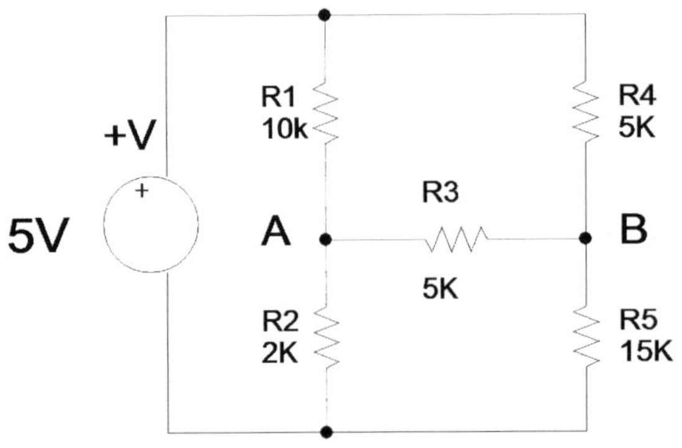

For this example we will consider R3 to be the load resistance (sensor, or whatever) and our efforts will be to determine the voltages at A and B, and thus the voltage across R3.

First, remove the load (R3) from the circuit. The circuit will then look like this:

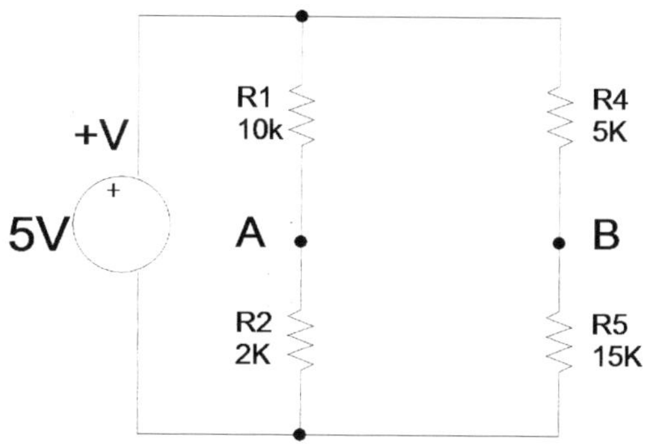

Then find the (Thevenin equivalent) voltage values at 'A' and 'B'.
With 5Volts applied, these will be:

$\frac{5}{12K} * 2K$ or 0.8333 Volts at point 'A' and

$\frac{5}{20K} * 15K$ or 3.75 Volts at point 'B'

So the Thevenin equivalent voltage across A and B is 3.75 – 0.8333 Volts or 2.91667 Volts. Now to determine the total resistance looking into terminals 'A' and 'B', with the voltage source (5V) shorted. The circuit then becomes;

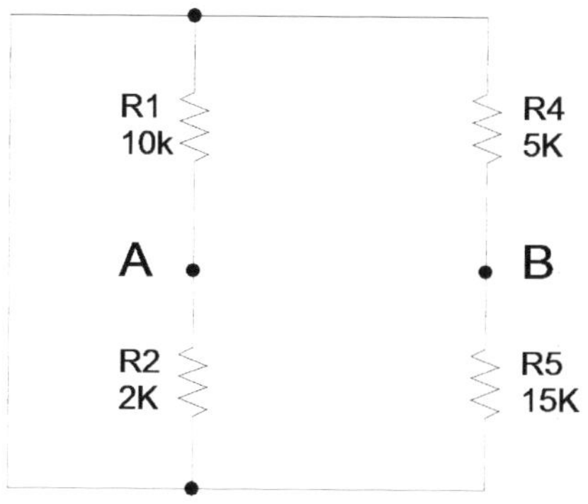

We will now have to look into 'A' and 'B' in order to determine the Thevenin resistance.

Re-drawing the circuit:

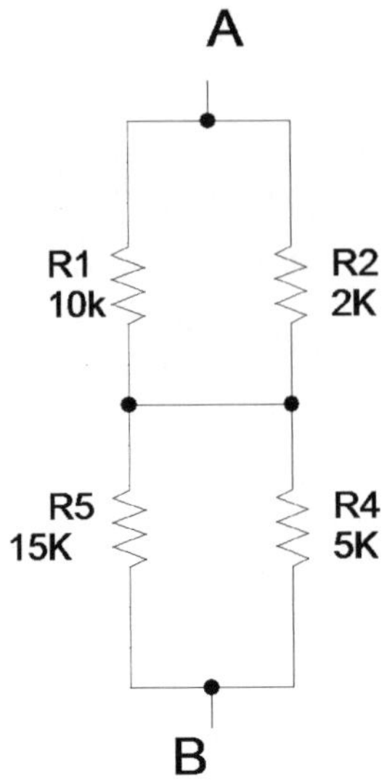

Take a close look at the circuit above. This is the whole key to Thevenizing a circuit component. The circuit above is the result of LOOKING INTO the 'A' and 'B' terminals (with the voltage source shorted). The calculation for the Thevenin resistance is:

$$\frac{10K * 2K}{10K + 2K} + \frac{15K * 5K}{15K + 5K} = 5{,}416.6667 \text{ Ohms}$$

So we set up the Thevenin equivalent circuit to determine the current through R3:

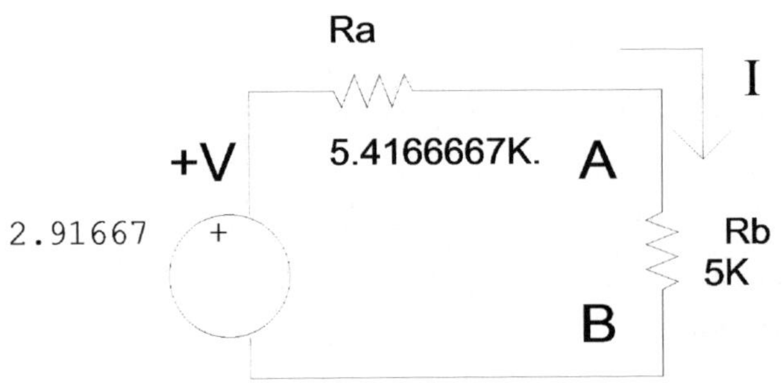

The current through the circuit is; $\dfrac{2.91667}{(5.4166667K + 5K)} = 0.28$ milliAmps.

The voltage drop across 'Rb' (5K resistor) is 0.28milli-Amps * 5K, or 1.4 Volts. So the voltage across R3 is 1.4 Volts.

We're halfway home! We now have the voltage across the load resistance but we need to know the potential, to ground at one end of the load resistance or the other. We'll do this by using a Thevenin equivalent of one of the resistors connected to 'A' or 'B'. For our example, let's select R2, the 2KOhms resistor. Remember, this is a totally arbitrary selection. Any of the four remaining resistors will do just fine. Remove R2, the 2KOhm resistor, from the circuit:

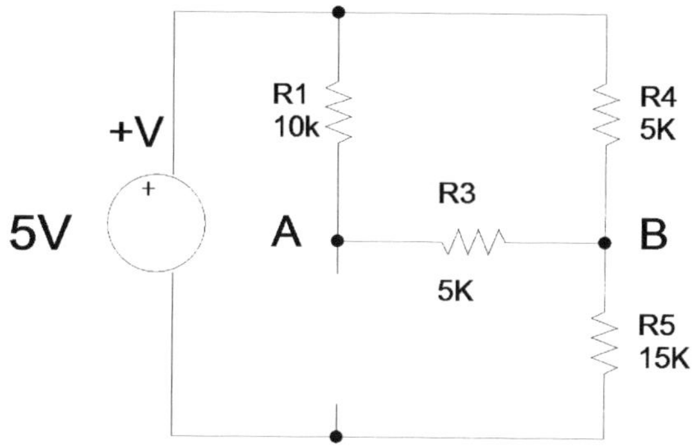

Calculate the voltage between 'A' and Ground. Calculating A we need to know the current through the entire circuit. The total resistance of this circuit is:

10K plus 5K in parallel with 5K, all in series with 15K or:

$$\frac{15K * 5K}{20K} + 15K = 18{,}750 \text{ Ohms}$$

The current through this circuit, with a 5 Volt source is:

$$\frac{5V}{18{,}750 \text{ Ohms}} = 2.66667^{-4} \text{ Amps} \quad \text{or } 0.2666 \text{ milliAmps}$$

So the voltage across R5 is:

$$2.66667^{-4} \text{ Amps} * 15K = 4 \text{ Volts}$$

Point 'A' is 1/3rd of the way between 4 Volts and 5 Volts, or 4.33333 Volts. Now to find the Thevenin resistance as viewed from points 'A' and Ground.

Re-drawing the circuit we get:

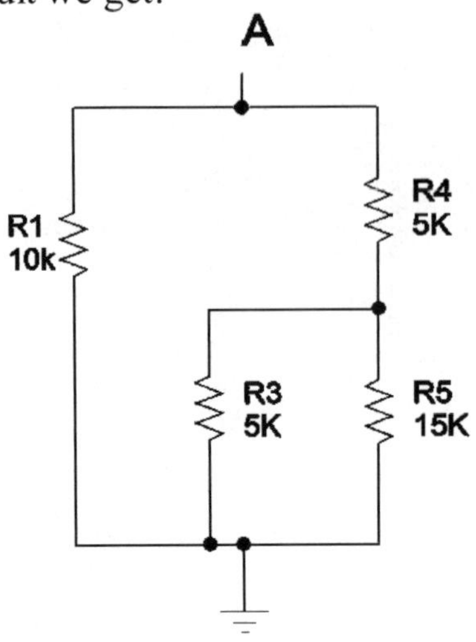

5K in parallel with 15K that is in series with 5K, all of which is in parallel with 10K or:

$$\frac{\left[\left(\dfrac{5K * 15K}{5K + 15K}\right) + 5K\right] * 10K}{\left[\left(\dfrac{5K * 15K}{5K + 15K}\right) + 5K\right] + 10K} = 4666.667 \text{ Ohms}$$

So, the Thevenin equivalent circuit becomes:

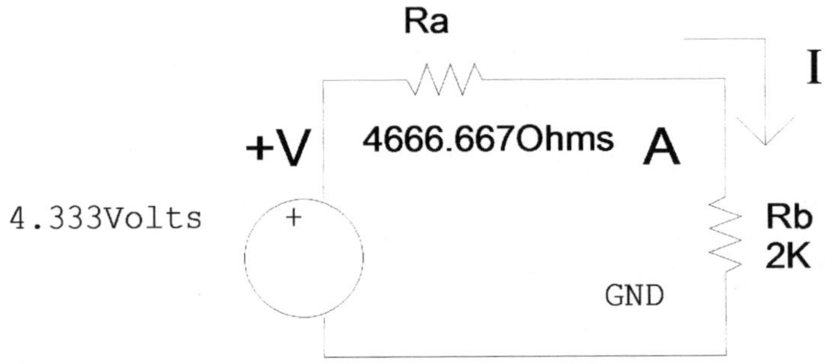

The current, I equals 4.3333V / (4666.667 Ohms + 2000Ohms) , or 650 micro-Amps. so the drop across the 2K resistor (R2) is: 650uA * 2K, or 1.3 Volts.

With ‘A’ equal to 1.3 Volts and a voltage drop across R3 of 1.4 Volts, then the voltage at ‘B’ must be 2.7 Volts above Ground. Let’s confirm these Voltage values by checking the current at nodes ‘A’ and ‘B’:

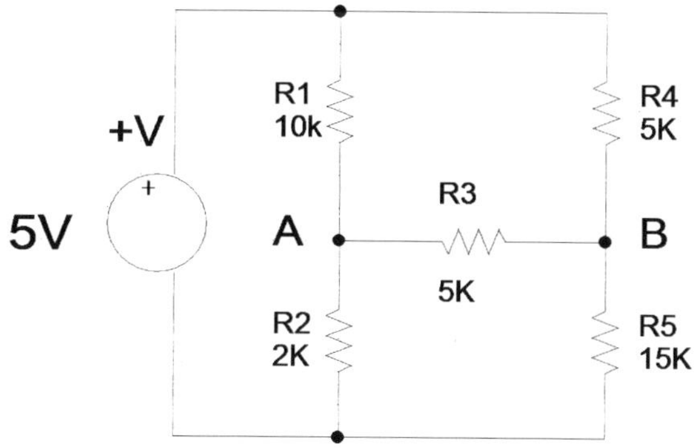

1. The voltage across R1 is 5V – 1.3V or 3.7V. The current through R1 is
   3.7 / 10K = 370 micro-Amps
2. With 1.3V at ‘A’, the current through R2 is;
   1.3/2K = 650 micro-Amps

3. The voltage across R3 is 1.4V so the current through R3 is
   1.4 / 5K = 280 micro-Amps.
4. The voltage across R4 is: 5V - (1.3V + 1.4V), or 2.3V.
   The current through R4 is: 2.3V / 5K = 460 micro-Amps.

5. The voltage across R5 is: 2.7V.
   The current through R5 is: 2.7V / 15K = 180 micro-Amps.

Checking:

a) I (R1) + I (R3) - I (R2) = 0

370uA + 280uA - 650uA = 0 Checks O.K.

b) I (R4) – I (R3) - I (R5) = 0

460uA - 280uA - 180uA = 0 Checks O.K.

So the node currents are in balance.

Enjoy!

Capacitance Independent Frequency-to-Voltage Converter

The following is a circuit that is used for converting Frequency to Voltage. The circuit elements are:

* Frequency input circuitry (pulse forming circuitry not shown)
* A constant current source
* Solid state switches used to gate the constant current either to an integrator circuit pseudo-GND or to GND.
* An integrator circuit with discharge resistor across the integrating capacitor.

Figure 1: Precision Voltage Source, Current Source and Switching Circuitry

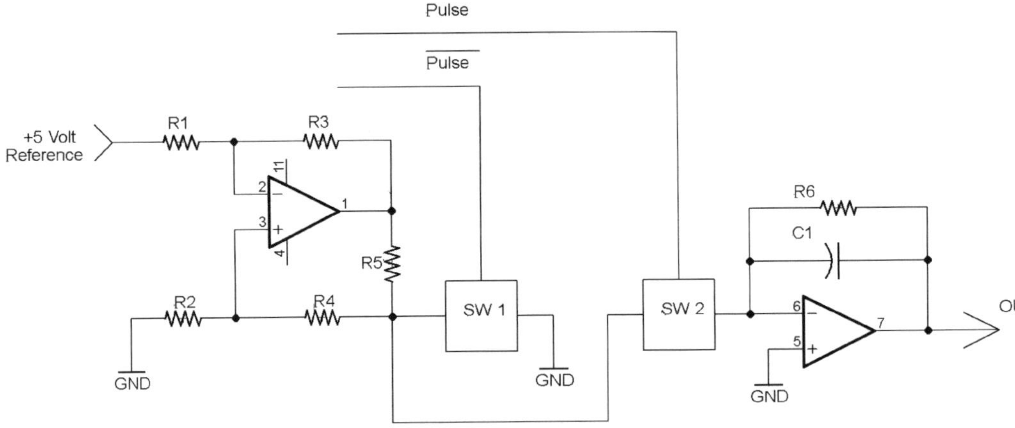

The input Frequency is used to produce pulses (circuitry not shown) with one pulse for every cycle of the frequency. In all of the following examples the period of these pulses has been chosen to be 10 microseconds (10 $\exp^{-6}$ sec.). Since the frequency will vary, the repetition rate of these pulses varies with the input frequency. Both the pulse itself and the time between pulses are used to control electronic switches (SW1 and SW2). These switches control the charging of an integrating capacitor (C1) as well as allow the discharging of this capacitor through a resistor (R6). In the circuit diagram, these switches are labeled as SW1 and SW2 (see Figure 1). The "Pulse" and "Pulse Not" signals are the inverse of each other. The "Pulse" signal is used to charge capacitor C1 with a constant current produced by the 'Improved Howland Current Pump' circuit made up of the operational amplifier to the left of SW1 and resistors R1 through R5. When the "Pulse" signal goes low, SW2 opens. Since the input to an operational amplifier usually has a very high impedance, the only discharge path for C1 is through R6. The operational amplifier is trying to maintain a 'ground' (GND) potential on pin 6. This is because pin 5 is at GND potential and the operational

amplifier is only satisfied when the voltages on pins 5 and 6 are the same. The output of this operational amplifier will follow the RC discharge voltage in an attempt to keep the voltages on pins 5 and 6 the same.
The"Pulse Not" signal is used to ground the output of the constant current source through SW1, while C1 is discharging. When the "Pulse Not" signal goes away SW1 will open and the "Pulse" signal will cause SW2 to conduct. With pin 6 of the operational amplifier trying to be at GND potential, the Current Pump circuit will see the same load as when it was grounded through SW1 to GND.

At first glance it appears that the accuracy of the circuit of Figure 1 is greatly affected by the value of the capacitor, C1. The following descriptive analysis of circuit operation includes how much affect input frequency changes and capacitor value changes have on the output of this Frequency to Voltage converter circuit.

Two formulae are required in order to define the integrator circuit and discharge circuits. These are:

$$C\,V = I\,T$$

and

$$\text{Percent of Discharge} = 1 - e^{-t/RC}$$

The first formula refers to the value of capacitor (C), the voltage (V) across this capacitor, the current (I) that charges C1 and the amount of Time (T) that this current is charging C1. The second formula defines the percent discharge of the capacitor (C) as it discharges exponentially through resistor (R).

The following values will be used:

C = 0.1uF (C1)
I = 2.9 milliAmps (out of the constant current source)
T = 10 micro-second (10 exp^{-6}) pulse width, the charge time of C.
t = Pulse Repetition Rate minus 'T', the discharge time of C.
R = 19.1kOhms (R6)

Note: Calculator math was used in these calculations. The number of decimal places in the calculated results do not conform to mathematical convention, but are used to attain more accurate results. Accuracy is especially needed when evaluating integrals.

Calculate the Voltage value (V) to which C will charge when a constant current of 2.9 milliAmps is applied to the capacitor for 10 micro-seconds:

$$V = IT/C \quad \text{or} \quad \frac{(2.9\ exp^{-3}\ \ x\ \ 10\ exp^{-6})}{0.1\ exp^{-6}}$$

$$V = 0.29\ \text{Volts}$$

<u>Assume an input frequency of 4000Hz</u>:
10usec charging pulses will occur at a Pulse Repetition Rate of 4000 pulses per second. The period from the start of a pulse to the start of the next pulse will then be 1/4000 or 2.5e -4 seconds.

The second formula is the general formula for the percent discharge of an RC network:

$$\text{Percent of Discharge} = 1 - e^{-t/RC}$$

From the information given above, "t" will be the Pulse Repetition Rate of $2.5e^{-4}$ seconds minus 10 microseconds, or $2.4e^{-4}$ seconds. The product of R and C (C1 and R6) is:

$$19.1\text{KOhms} * 0.1\ exp^{-6} \quad \text{or} \quad 1.91\ exp^{-3}.$$

$$\text{Therefore, the Percent of Discharge of C1} = 1 - e^{[(-2.4\ exp\ -4)/(1.91\ exp\ -3)]}$$

$$= 0.118080458\%$$

Between charge pulses of current, C1 will discharge through R6. It will discharge for $2.4e^{-4}$ seconds, until the next pulse of current arrives. From the above calculation, it will discharge 0.1180805 percent of the total Voltage applied across it. During the first pulse of current it will charge to 0.29 Volts (see previous calculation using CV = IT). It will then discharge until the voltage is "0.29 Volts minus (0.29 * 0.11808%)". This value is 0.29 - 0.034243 or 0.255757 Volts. When the next pulse arrives, the voltage will be increased to 0.2557 Volts + 0.29 Volts or 0.5457 Volts. Multiplying this voltage by 0.1180805% gives us 0.064443 Volts. Subtracting this value from 0.5457 Volts gives us 0.5457 – 0.064443 or 0.48125 Volts, the voltage to which C1 will discharge after the second current pulse. Notice that the amount of voltage being discharged is increasing. The voltage discharged after the first pulse is 0.29 – 0.2557 = 0.034243 Volts while the voltage discharged during the second pulse is 0.54570 minus 0.48125 = 0.06444 Volts.

If we continue to put 10 microsecond wide pulses of current into C1 the voltage across C1 will continue to increase until the Voltage value by which it increases with each pulse (0.29 Volts) equals the amount of voltage that is discharged from C1 through R1 during the 2.4exp-4 discharge period. This only occurs when 0.1180805 percent of the voltage across C1 equals 0.29 Volts. The formula below mathematically represents this situation of equilibrium.

$$1 - [\text{ Volts across C1 } / (\text{Volts across C1} + 0.29 \text{ Volts})] = 1 - e^{[(-2.4 \exp -4)/(1.91 \exp -3)]}$$

Simplifying:

$$\text{Volts across C1}/(\text{Volts across C1} + 0.29 \text{ Volts}) = e^{[(-2.4 \exp -4)/(1.91 \exp -3)]}$$

$$\text{where} \quad e^{[(-2.4 \exp -4)/(1.91 \exp -3)]} = 0.881919542$$

Cross Multiplying:

$$\text{Volts across C1} = 0.881919542(\text{Volts across C1}) + 0.881919542(0.29 \text{ Volts})$$

Subtracting:

$$0.118080458(\text{Volts across C1}) = 0.255756667 \text{ Volts}$$

Solving:

$$\text{Volts across C1} = 2.165952517 \text{ Volts}$$

This is the voltage AFTER C1 has discharged to 88.19%. So the peak voltage is 2.165952517 + 0.29 or 2.455952517 Volts.

To get the average voltage as C1 discharges we need to integrate the area under the exponential discharge curve. This will give us a more accurate voltage reading.

$$\text{Area} = 2.455952517 \int_{o}^{2.4\exp^{-4}} e^{[-1/(1.91 \exp -3)]t}$$

From any table of integrals: $\int e^{at}$ is $\frac{e^{at}}{a}$

Evaluating the integral, the area is:

$$= 2.455952517 * \left[\frac{e^{[-1/(1.91 \exp -3)]t}}{-1/1.91\exp^{-3}} \Bigg|_{0}^{2.4 \exp^{-4}} \right]$$

$$= 2.455952517 * \left[\left(\frac{0.881919541}{-523.5602094} \right) - (-1.91 \exp^{-3}) \right]$$

$$= 2.455952517 * [\; 2.255336766 \exp^{-4} \;]$$

$$= 5.3900000 \exp^{-4}$$

Evaluating this integral gave us an area under the voltage curve of **5.539 $\exp^{-4}$**:

Figure 2: Waveform when the Percent of Discharge is equal to 0.118080458 (or 11.808%).

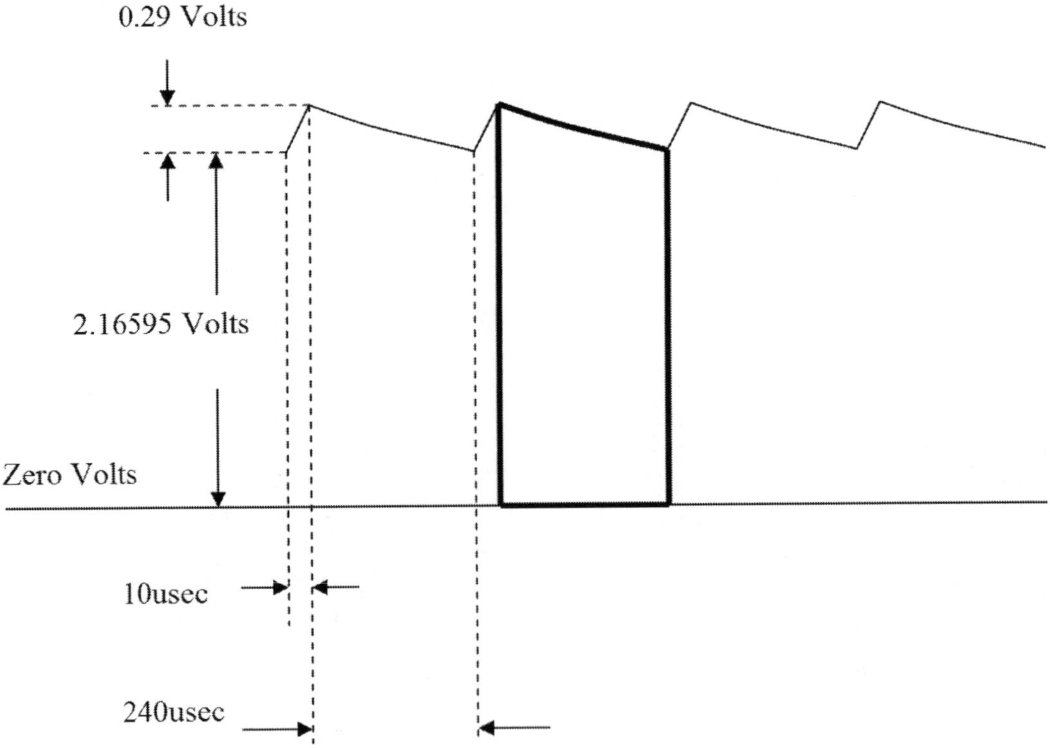

So what happens when the value of the capacitance changes?

Let's reduce the capacitance value by an order of magnitude while leaving the input frequency the same (4kHz):

C = 0.01uF
R = 19.1KOhms
I = 2.9milli-amps
T = 10 micro-seconds

The new Voltage that is added to C1 with each current pulse is now:

$$V = IT/C$$

$$= \frac{(2.9\ \exp^{-3})(10\ \exp^{-6})}{0.01\ \exp^{-6}}$$

= 2.9 Volts per current pulse

The RC product will be 19.1kOhm * 0.01uF or $1.91\exp^{-4}$

Therefore, the Percent Discharge of C1 = $1\text{-}e^{[(-2.4\ \exp\ -4)\ /(1.91\ \exp\ -4)]}$ = 0.715364112

The equation for finding the point at which the percent discharge in voltage equals the increase in voltage due to a charge pulse becomes:

1 - [Volts Across C1 / (Volts Across C1 + 2.9 Volts)] = $1 - e^{[(-2.4\ \exp\ -4)/(1.91\ \exp\ -4)]}$

Simplifying:

[Volts Across C1 / (Volts Across C1 + 2.9 Volts)] = $e^{[(-2.4\ \exp\ -4)/(1.91\ \exp\ -4)]}$

where $e^{[(-2.4\ \exp\ -4)/(1.91\ \exp\ -4)]}$ = 0.284635887

Cross Multiplying:

Volts Across C1 = (0.284635887 * Volts Across C1) + (2.9)(0.284635887)

Subtracting:

0.715364112(Volts Across C1) = 0.825444074

Solving:

Volts Across C1 = 1.153879626 Volts

Remember, this is the voltage AFTER C1 has discharged to 0.715364113%.

So, the peak voltage is actually 1.153879626 Volts + 2.9 Volts or 4.053879626 Volts. To get the average voltage for one period we need to integrate the area under the voltage curve as C1 discharges during the 2.4 $\exp^{-4}$ seconds time period:

$$\text{Area} = 4.053879626 \int_{0}^{2.4 \exp^{-4}} e^{[-1/(1.91 \exp -4)]t}$$

Once again, the integral of e^{at} is $\frac{e^{at}}{a}$

Evaluating the integral, the area becomes:

$$= 4.053879626 * \left[\frac{e^{[-1/(1.91 \exp -4)]t}}{-1/1.91\exp^{-4}} \Bigg|_{0}^{2.4 \exp^{-4}} \right]$$

The resulting area under the discharge voltage curve is **5.539 exp -4**:

Figure 3: Waveform when Percent of Discharge of C1 is equal to 0.715364113 (71.536%)

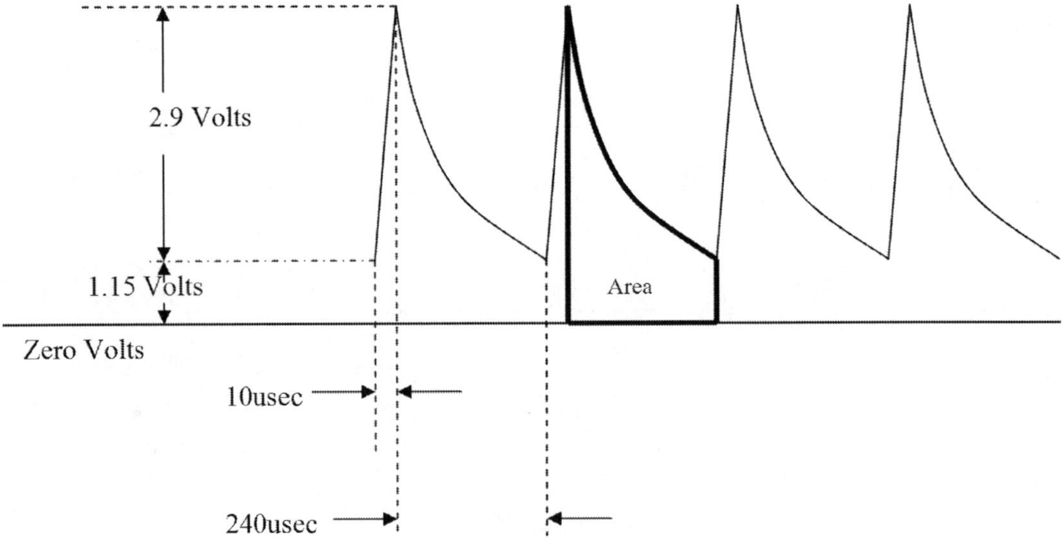

How about that!
The area under the capacitor discharge curve is exactly the same as for a 0.1uF capacitor. The peak voltage, and therefore the amount of ripple on the signal, increases significantly but the area under the voltage curve during discharge remains the same.
Increasing the capacitance value to 1uF will results in the same area under the curve, although the work is not shown here.

So, in this F/V converter circuit, when a fixed width pulse of constant current is applied to the capacitor charging circuit, the peak voltage to which the capacitor (C1) charges may change based upon the value of the capacitor. But this change is offset by the change in the rate of discharge of the capacitor through the resistance (R6).

Well! That sure was interesting.

Now let's find out what happens when the frequency is changed.

<u>Keeping the original values and changing the input frequency to 2kHz</u>:

Period = 1/frequency = 1/2000 = 500usec
C = 0.1uF (C1)
I = 2.9 milliAmps (out of the constant current source)
T = 10 micro-second ($10e^{-6}$) the pulse of charge for C
t = 490usec (500usec – T) the discharge time for C.
R = 19.1kOhms (R6)

As in the first and second examples, the same value of voltage increase will be attained with each pulse of current, i.e.

$$V = IT/C \quad \text{or} \quad \frac{(2.9\text{exp}^{-3} \times 10\text{exp}^{-6})}{0.1\text{exp}^{-6}} = 0.29 \text{ Volts}$$

The value of the RC product is still 19.1KOhms * 0.1exp^{-6} or 1.91exp^{-3}.

So, using the new value of 't', the Percent of Discharge becomes:

$$\text{Percent of Discharge} = 1 - e^{-t/RC}$$

$$= 1 - e^{[(-4.9 \text{ exp } -4)/(1.91 \text{ exp } -3)]}, = 1 - 0.773720561$$

$$= 0.226279438$$

(or 22.6279438%)

The equation for finding the point at which the percent discharge voltage equals the increase in voltage due to a charge pulse becomes:

$$1-[\text{Voltage Across C1}/(\text{Voltage Across C1}+0.29 \text{ Volts})] = 1 - e^{(-4.9 \exp -4)/(1.91 \exp -3)}$$

Solving for "Voltage Across C1" (see above for solution steps) we get a value of 0.991601199 Volts. Since this is the voltage remaining AFTER C1 has discharged by 22.6279438%, the peak voltage is actually (0.991601199 Volts + 0.29 Volts) or 1.281601199 Volts .

Calculating the average voltage for one period:

$$\text{Area} = 1.281601199 \int_{0}^{4.9 \exp^{-4}} e^{[-1/(1.91 \exp -3)]t}$$

Once again, the integral of e^{at} is $\frac{e^{at}}{a}$

Evaluating the integral:

$$= 1.281601199 * \left[\frac{e^{[-1/(1.91 \exp -3)]t}}{-1/1.91\exp^{-3}} \Bigg|_{0}^{4.9 \exp^{-4}} \right]$$

We get an area under the voltage curve of **5.539 exp^{-4}**

Wow! The PEAK voltage changes from 2.45595 Volts for the 4kHz input frequency to 1.2816 Volts for the 2kHz input frequency and the area under the voltage discharge curve remains the same.

We will need a few more points in order to determine if there is a solid relationships between either peak voltage or area under the discharge voltage curve as the frequency is changed. We already know the results for 4kHz and 2kHz when the original component values are used, so lets find and tabulate the results for a range of 1kHz to 5kHz, in 1kHz increments. Using the original component values, current and current pulse period in a spread-sheet program, the resulting values are:

Frequency	t (seconds)	RC Product	e(-t/RC)	1-e(-t/RC)	Peak Voltage	Integrated (V/t)Value
5khz	-1.90000E-04	1.91000E-03	0.905311279	0.094688721	3.062666775	5.53900E-04
4khz	-2.40000E-04	1.91000E-03	0.881919541	0.118080459	2.455952517	5.53900E-04
3khz	-3.23333E-04	1.91000E-03	0.844268717	0.155731283	1.862182048	5.53900E-04
2khz	-4.90000E-04	1.91000E-03	0.773720561	0.226279439	1.281601199	5.53900E-04
1khz	-9.90000E-04	1.91000E-03	0.595517438	0.404482562	0.716965396	5.53900E-04

Results:
The area under the discharge voltage curve doesn't vary with changes in frequency.

Graphing the Peak Voltage vs. Frequency we get:

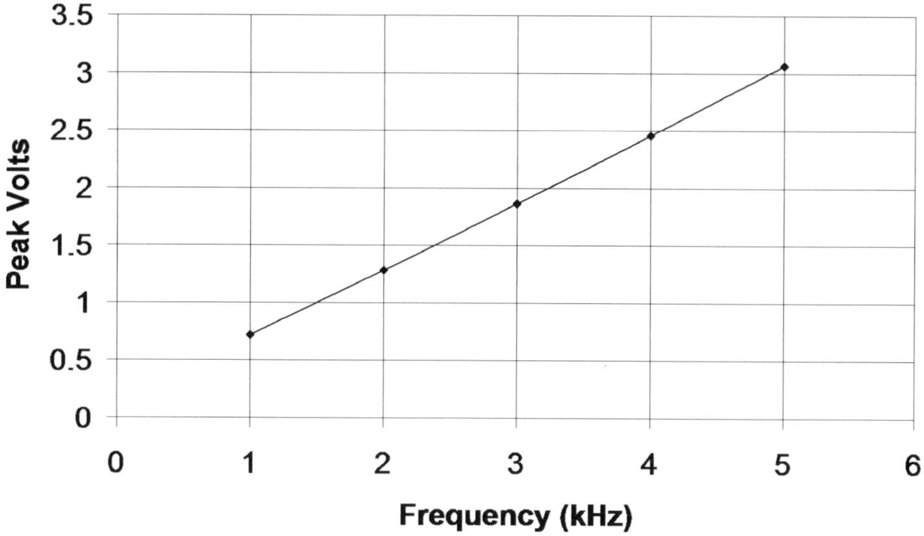

Looks pretty linear.

But let's draw a straight line between the 1kHz voltage and the 5kHz voltage:

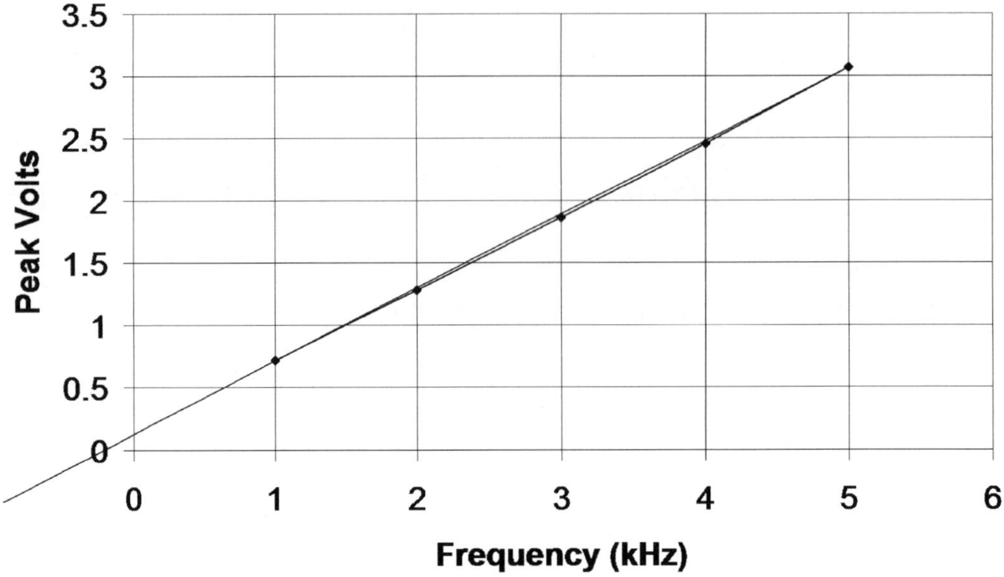

Uh oh! Looks like the Peak Voltage sags a little.
You may want to check the behavior of this circuit at frequencies below 1kHz because the minimum Peak Voltage level should be 0.29V (the voltage increase for one pulse of current) and this graph shows it as a lot lower than 0.29V when approaching 0 Hz. O.K., o.k! I did it already (I couldn't stand not knowing. It's like a good mystery novel). Here's the graph with peak voltage values at 100Hz, 200Hz, and 500 Hz added:

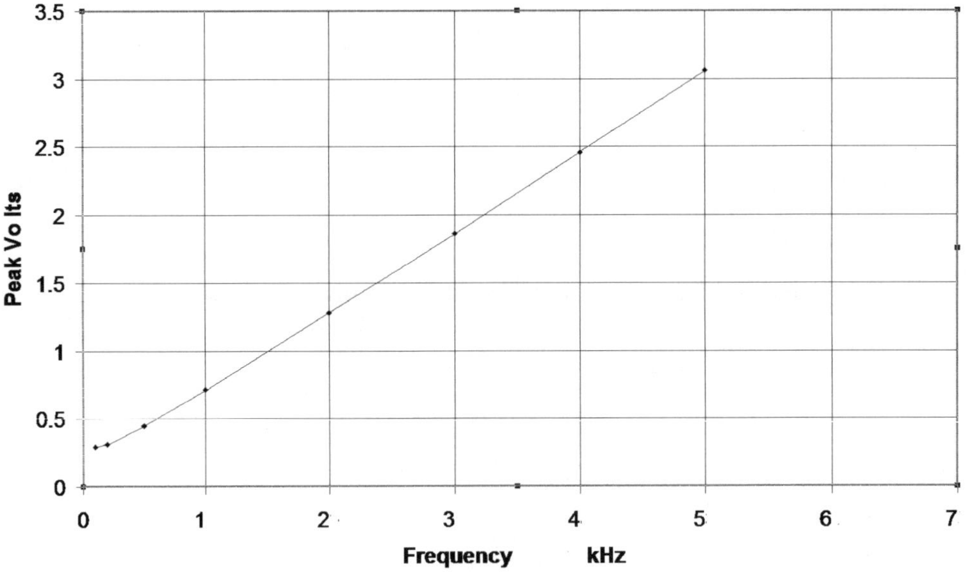

So, there is definitely some distortion of the F/V conversion at the lower frequencies. It also appears that the Peak Volts may not go lower than 0.29 Volts. That would make sense. One period of time that is interesting to note is the amount of time between the initial application of the frequency input and the point at which the output voltage from the F/V circuit is stable. A spreadsheet was used to calculate this time for various values of capacitance (C1) in the circuit of Figure 2. The times that are given in the results that follow represent the times from zero charge on the capacitor to full charge on the capacitor at a frequency of 4kHz.

For a 0.01 uF capacitor the response time is 3.5 milliseconds.
For a 0.1uF capacitor the response time is 29 milliseconds.
For a 1.0 uF capacitor response time is 265 milliseconds.

Well, I think we've beaten this dead horse just about enough.

Keep thinking.

Things to Consider when applying Flexible Automation (Robots)

It wasn't too many years ago that Robots were the word of the day, especially in U.S. Industry. They called it 'flexible automation'. Industry always seems to be jumping on some band-wagon. I've been through the Just in Time, Lean, ISO-9000, Outsourcing and Flexible Automation band-wagons. But I digress.

Robots are made with varying degrees of mechanical flexibility (articulation) that ranges from two axes to six or more axes of motion (degrees of freedom). Robots are usually either electric or hydraulic, although a few robots operate on compressed air. The electric robots usually use D.C. motors. The reliability of robots has increased somewhat over the past 25 years.

A robot itself is deaf, dumb and blind. It depends upon sensory inputs such as ultrasonic transducers, interruptible light beams, proximity sensors, switches, visual interpretation devices and tactile sensors to tell the robot when it is within the proper operating parameters and when the materials to be handled are in position. If either the robot material handler or the material being handled is not in correct position, damage can be done to the robot. Hitting a cardboard box edge-on is almost the same as running into a brick wall. In many manufacturing operations the cost of sensors can meet or exceed the cost of the robot. The key to any robot installation also depends a great deal on the design of the "end effector". The end effector is the business end of the robot. For example, in a material handling application it is the gripper and in a painting application it is the spray applicator assembly. End effectors are usually a separate design from the robot itself. The robot just gets you there. It's up to you to do something after you're there.

The most successful applications for robots are non-contact applications. These include spray painting, application of viscous materials such as bonding materials and the application of chemicals. Welding is a contact operation, but the load is very light and almost constant. The majority of robot applications are still in welding and spray painting applications. However, robots are useful in environments that are inhospitable to humans such as high or low temperature areas and in gaseous or explosive atmospheres. A robot application may be worth the time and expense in situations that are harmful to humans. The application must demand a high degree of flexibility, otherwise hard automation (dedicated machines) will probably be a better solution.

The original intention of applying robots was to reduce the labor force. Believe me, even the simplest task for a human can be extremely difficult, if not unachievable when applying a robot to the task. There are no "universal" applications. Each application involves its own set of problems. This means that the expense of each new application will be as high as the last. Don't get me wrong! Robotics is the way to go, it just has a long way to go.

Some General Rules to consider when applying robots:

- Always weigh dedicated or "hard" automation against the flexibility of a robot. The cost of this flexibility is high and, in many cases, the problem has already been solved using hard automation.

- Don't fall into the trap that robots are truly "flexible automation" and can easily be transported from task to task. Each task is unique and requires a great deal of planning and support hardware.

- Look for simple applications involving operations requiring a minimum of dexterity but involving three or more degrees of freedom and several motions. Applications involving repetitive point-to-point motions can usually be solved with less expensive hard automation.

- Look for application that are difficult for people to perform, such as applications involving high temperature product, caustic chemicals or non-life supporting atmospheres. Once again, be sure that the task merits flexible automation rather than hard automation.

- Consider the total cost of the application. The robot may turn out to be the least expensive part of the project.

- Consider the usage of the robot system (robot and associated hardware and sensing). Unless there are unusual circumstance to the application, continuous duty over two or three shifts should be a requirement. Also, consider maintenance in any "payback" calculations.

That's about it.

Digital Logic in Analog Applications

So here's a novel twist. Let's see how digital logic works as an analog device. Monostable devices are normally considered to be more analog than digital. They are normally 'programmed' by using capacitors and/or resistors. Well guess what. Here are a few circuits that can lengthen, shorten or delay signal edges using diodes, resistors and capacitors. These circuits also depend upon Schmitt trigger CMOS devices and their switching points. Normally, the switching point for a HIGH level is anything above 2/3 of the power supply voltage (Vcc), and the switching point for a LOW level is anything below 1/3 Vcc.

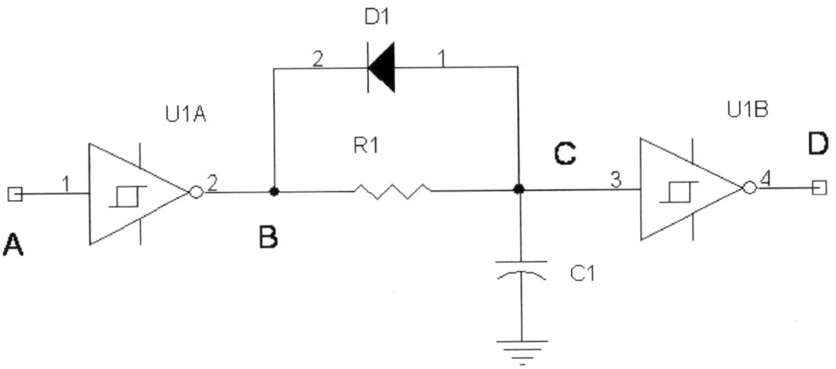

The circuit above is used to delay the falling edge of the input signal. It can also be considered as a 'positive pulse stretcher'. The core device is a Schmitt Trigger CMOS logic inverter. Let's say that the signal at 'A' has just fallen to a low level. This means that the output of U1A or the signal at 'B' rises to a high level. This begins to charge C1 through R1. As C1 is charging, The voltage at 'C' passes through the switching point of U1B causing point D to go negative. The time constant of R1/C1 determines how long it takes to get to this switching point. When the signal at 'A' goes positive, the output of U1A goes negative. D1 is then forward biased and gives C1 a path of discharge. This discharge path is from 'C', through U1A power supply, through D1 and back to C1. C1 discharges rapidly in this direction.

The RC time constant of R1 and C1 determines the delay time of this circuit. This is a ratio of R to C. When the value of C is kept small and the value of R is kept large but the ratio of R to C is kept constant it is easier to discharge C without having to use a lot of drive. Actually, U1A can't produce a lot of drive anyway, but U1A and U1B have high input impedances and will not load down the RC circuit. So try to keep R large and C small while maintaining the RC time constant you need for the proper delay. So what do the waveforms look like at B, C and D with a square wave input at A?

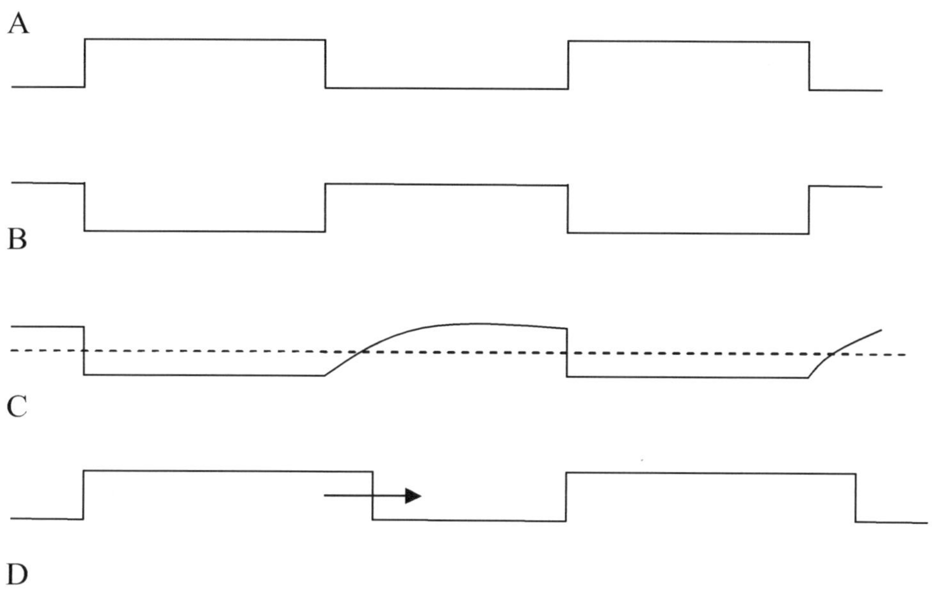

The length of the positive going pulse is 'stretched'.

Now let's move a few components around and see what happens.

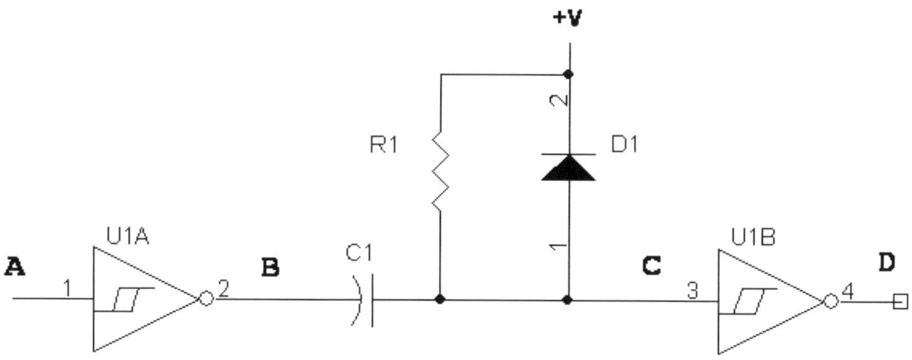

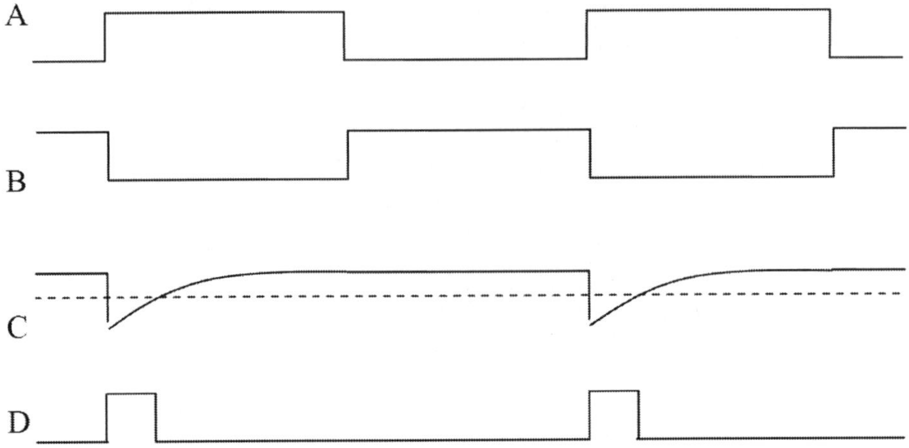

So this turns out to be a pulse shortener or 'squeezer'.

How do we figure out the length of the stretch or squeeze? We have to lay down some ground rules:

1. First is that the output of the CMOS logic ships actually goes from ground to +V.
2. The second is to remember that the switching points of the Schmitt Trigger CMOS logic chip inputs are as follows:
   a) Any input lower than 1/3 of +V is considered to be a Low level
   b) Any input higher than 2/3 of +V is considered to be a High level.
3. The third thing is to set the amount of time that you want the signal delayed. Let's make the delay time be 3 milliseconds and lets select the pulse stretch circuit. This means that the input pulse (low-to-high transition then high-to-low) will be lengthened by 3 milliseconds.
4. The fourth is to set some value for +V. In this case, let's set a value of +6Volts.

We really don't care about the width of the incoming pulse. Whatever it is, it will be increased by 3 milliseconds. Here we go:

In the first circuit, the threshold (dashed line) occurs at 2/3 of +V or 4 Volts. The RC charge time is represented by the formula:

% of +V = e ^ (-t/RC)

We know that the % of +V at the switching point (2/3 of +V) is 0.66667.

So the formula becomes:

0.66667 = +V * e ^ (-t/RC)

Dividing 0.66667 by +V we get 0.13333

Taking the natural log of both sides:

ln (0.13333) = -t/RC

or -2.015 = -t/RC

We know that the value of 't' is 0.003 seconds (3 milliseconds), so the product of R and C has to be t / 2.0150 or 1.4888exp-3. This is approximately 0.001488.

Let's select a capacitor value first. A value of 0.1uF sounds as good as any so the value of R has to be:

R = 0.001488 / 0.1 exp-6 or 14,880 Ohms (approximately 15KOhms).

The time the calculation is made in the same way for the pulse width 'squeeze' circuit, but the result is the width of the output pulse starting from when the input ('A') goes high.

Hooray for CMOS!

Traveling for Your Work

I remember my first trip for an employer. It was a coast-to-coast, one week trip for training. The coast I left was New York, the coast I went to was 'silicon valley', California. That was a good trip. They had a mild earth quake while I was there but I was in a moving car so I didn't even notice it. If I had any trip to do over, it would probably be that one (I even got to see one of the first showings of 'StarWars').

The second trip was for a different employer. This one wasn't as much fun by a long shot! I had been with a small company for about four or five weeks. I was just getting a handle on the circuitry when the chief engineer scheduled me for a trip to a customer that owned a strip-mining operation in Kansas. I was just supposed to be a 'place holder' until the field engineer returned from another job. Anyway, I wound up sitting across the table from the Chief Field Engineer for Bucyrus-Erie, a manufacturer of very large equipment for the strip-mining industry, who kept asking me why my company's equipment wasn't working properly. After he discovered that my company had sent me out 'cold', he made a few phone calls. So did I. To make a long story short, I wound up staying one and one-half weeks and fixing some problems. I only mention it because it's memorable, to me.

There were a few more trips with this employer; an automotive glass manufacturer in Toledo Ohio, a transmission assembly plant near Ypsilanti Michigan and another trip to Kansas. All of these trips were of a troubleshooting nature, but they certainly helped me to understand the product. On all of these trips, the days were long. I would arrive at work at the same time as everyone else, but would remain at work until 8:00 P.M. or 9:00 P.M. every night in order to get the problem fixed. That turned out to be the norm for road trips. Lots of hours. The only consolation was a steak dinner every night, but even steak dinners wear thin when you're away from home. So how do I sum up road trips for troubleshooting? Routine breaking, family disrupting, long work day, lonely but highly educational and well fed times. That's about it.

On a par with trips for training is the trip for an industrial show. These are usually short trips involving good lunches and dinners and, possibly introductions to some new technologies. The only problem is that you're usually traveling with co-workers. Hey! If you were at work you'd be with them anyway. These shows are usually only every six months or a year. Some of them switch coasts every other year.

You may never have to go to a show or you may have to go to shows that have your products on display. It's a whole different flavor when you're showing your own product. This involves pre-show coordination back at your plant and usually, a mad rush as the show date approaches. You can often reason with customers when you have a delay but 'a show waits for no man'. So you may have put in a lot of hours before you even get to the show. Then there is equipment unpacking, set-up and de-bug. Now for the bad news. You may wind up on your feet for most of the show. You may have to double as a salesman in some cases, but usually you're there just in case someone has a technical question, in which case you have to dance around the explanation in order not to 'give away the farm'. The competition is usually the one asking the questions. After the show there is the packing up (a lot of cramming, stuffing and leaving non-essentials behind). Packing up usually comes with a deadline also, otherwise there are charges. So when you finally check out of your hotel and are on the way to the airport you're pretty well drained.

But all of it is memorable and a heck of an experience.

Uni and Bi Directional Series Current Limiter Circuits

When you need a circuit that limits current but must operate over a wide range of voltage (say 3 Volts to 50 Volts) then this may be just the ticket.

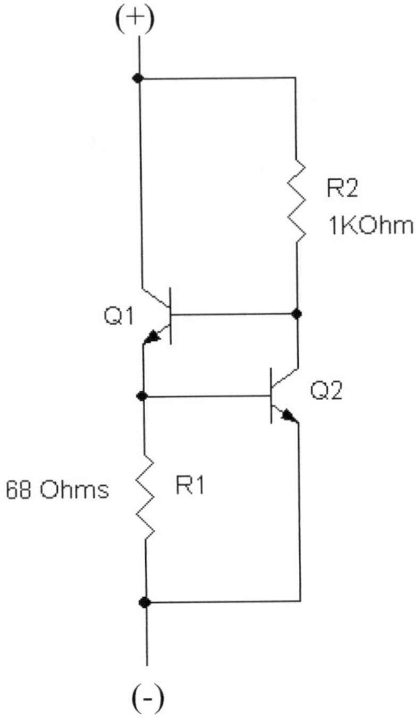

The whole thing depends upon the emitter-base drop of Q2. Q1 is just there to be turned off by Q2. This is basically the same set-up as the current limiting scheme on the "series-pass" linear voltage regulators, without the voltage regulation. Anyway, Q2 conducts when its emitter-base junction voltage is above approximately 0.7 volts. Hey, nobody said that the current regulation was going to be accurate. It's somewhere around +/- 5% of some point and varies all over the place with temperature. So what. When you need current limiting, you need current limiting.

How about limiting the current in both directions. The circuit below will do that using the same current sense resistor, R1. The current limit for each direction will be fairly close. It's just the circuit above, mirrored for current in the opposite direction.

An example of a polarity independent current limit circuit:

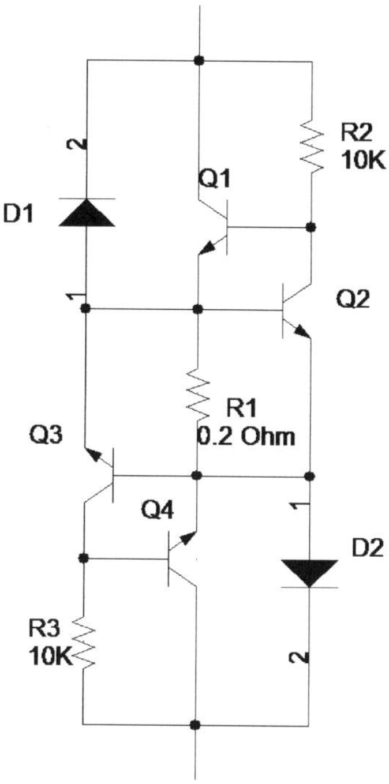

The whole idea is that this current limit circuit can be put in series with *any* line that needs current limit protection. The thing to remember is that Q1 and Q4 must be able to handle the circuit's maximum current and that Q1 and Q4 may wind up operating in their linear regions and dropping a lot of power. In these instances they will have to be heat sunk.

Onward and upward!

Hard Lessons – Get the Specs!

I can't emphasize this enough. Whenever you are designing a circuit "get the specs and check the specs" of all active and passive components you are using. There are many websites for gathering this data. When I first designed a high voltage circuit, I smoked some resistors. I wondered why? They were of the proper value and wattage. Then I found out that they have a voltage breakdown rating. They were breaking down. The solution was to series a couple of resistors with half the value. This way they split the voltage drop between them.

Another thing that can throw you is a fuse. These little things aren't just zero ohm devices that open immediately when some current value is exceeded. They are a rather dynamic device. They have a 'current versus time' value often referred to as an I2T (I squared T) value. I've seen fuses, which are close to their limit, light up like a light bulb and warp out of shape, then cool down and go back to their original shape time and time again without blowing. Don't be afraid to get a data sheet for a fuse. Actually, most fuse manufacturers have very detailed data or product sheets on their products (I think they expect you to use them).

The thing that can throw you the most is the data sheet of a new component, or one with which you are unfamiliar. Remember, the mistake is always yours. We all assume that 'they couldn't be that stupid'. Believe me, they can. I've run into components that work fine until you get to 90 degrees Fahrenheit ('Fahrenheit'!). This wasn't the published temperature spec but the manufacturer never wound up eating these parts, the company that I was working for ate them. I've run into op-amps that were advertised to be 'latch-up free' but after sitting with power ON for eight hours, would latch up every time. I've run into components that source current from their high impedance inputs. This turned out to be a spec that was, as I see it, made purposefully vague. (After they manufacture the chips they have to sell them too! After all, they're in business).

It doesn't matter if it's a digital component or an analog component, the drill is the same. Don't be afraid to ask the question. If you don't see a statement or a graph that indicates a parameter you are particularly interested in, give these guys a call. Once you design this POS (Piece of S%#@_) into your circuit, you'll have to live with it. Do the work up front.

You'll be happier for it.

Simplest Motor Speed Control Circuit

Here's about the simplest control loop you'll find. The whole thing consists of a mono-stable multivibrator, a drive component (transistor, in this case), a motor and an optical encoder (or tachometer. Anyway, something that generates pulses) It's shown with a re-triggerable one-shot (monostable), but a non-re-triggerable will do just fine (74121). CMOS logic is fine, and may help with noise immunity if operated at a higher voltage. Q1 is as hefty as you need it to be. It could be a power Darlington if drive isn't enough. No current protection is shown but I'd put it in if I were you.

Notice that the signal out of the monostable that drives the transistor is the inverting (Q Not) signal. This means that the time that the monostable is OFF determines how long the transistor is ON. Interesting, no?

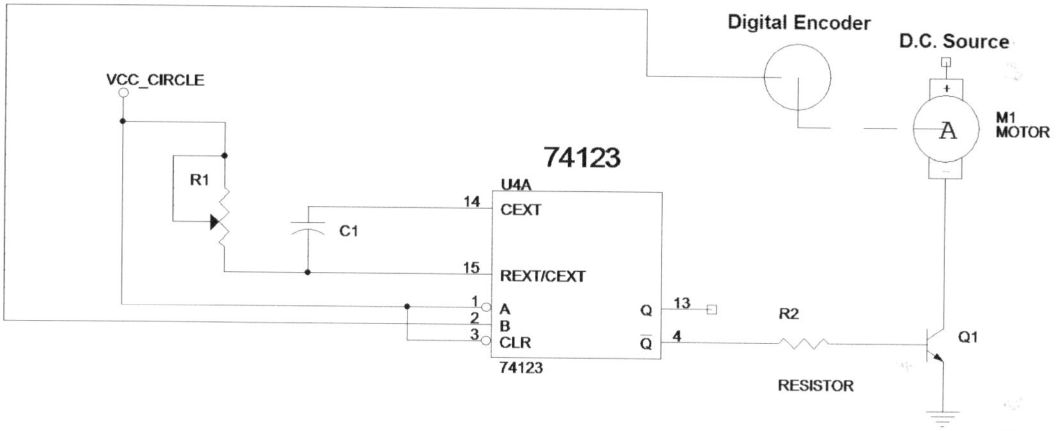

So the OFF time of the drive circuitry is really what is being controlled. When you increase the load, the system slows down. But this increases the OFF time from the tachometer and more power is applied to the motor. So the circuit winds up operating at one speed even with large changes in load. It's basically a cheap and dirty speed control
with R1 used to set the speed.

The time at which the speed is most affected is when the load is suddenly removed. Then the only load remaining is the motor load itself and the system may overspeed momentarily (depending upon the resolution of the encoder/tach). But, hey who wants perfection anyway?

It's difficult to calculate the resulting speed because the behavior of the motor has to be very well known. Some motor speed-torque calculations have to be included, along with the resolution of the tachometer (optical encoder) and the Resistor and Capacitor values.

Best of luck.

Traveling to Work

Let's face it! It is always nice to have a short commute. Fifteen to twenty minutes sounds just about right for being awake enough to face the work day. Well, that's getting harder and harder to accomplish. I have had commutes that ranged from seven minutes (or 25 minutes on a bicycle) to one hour and twenty seven minutes ONE WAY! All of this was done in an effort not to have to move my home.

I have already mentioned the reasons why you may lose a job ('fat trimming' lay-offs, mergers and acquisitions, selling the company, a new boss who has inherited you….). All of these things lie in wait for the employed engineer. As a side note, engineers are more susceptible to having these changes affect them because they just want to do their jobs. They are highly motivated toward a particular end, the creation of something new.

So what happens when you have to get a new job? In most cases it depends upon how good or bad the economy is. If the economy isn't too bad, you may be able to get a job fairly close to home. However, if the economy stinks, you may find yourself following up leads on jobs that aren't even in the same state. While living in Connecticut I have had all-expense-paid interviews in New York, Michigan, Florida, North Carolina, and Georgia to name a few. It seems as if I have always been offered a job within the State of Connecticut at the last minute and have wound up staying in the same school district for over 25 years. However, as I mentioned earlier I have wound up with commutes that put me in a car for three hours a day. Needless to say, my wife did all the driving on the weekends.

Engineers, and the engineering profession have become "Nomads of the 21st Century". Many have had to 'go where the jobs are', and many of these jobs have been on the other side of the U.S. For those who choose contract work this is getting to be the norm. But for engineers who want to stay in one place, they may have to use the 'knocking on doors' technique in order to remain in the same town for any length of time.

Keep up the good work.

Analog Delay Circuit

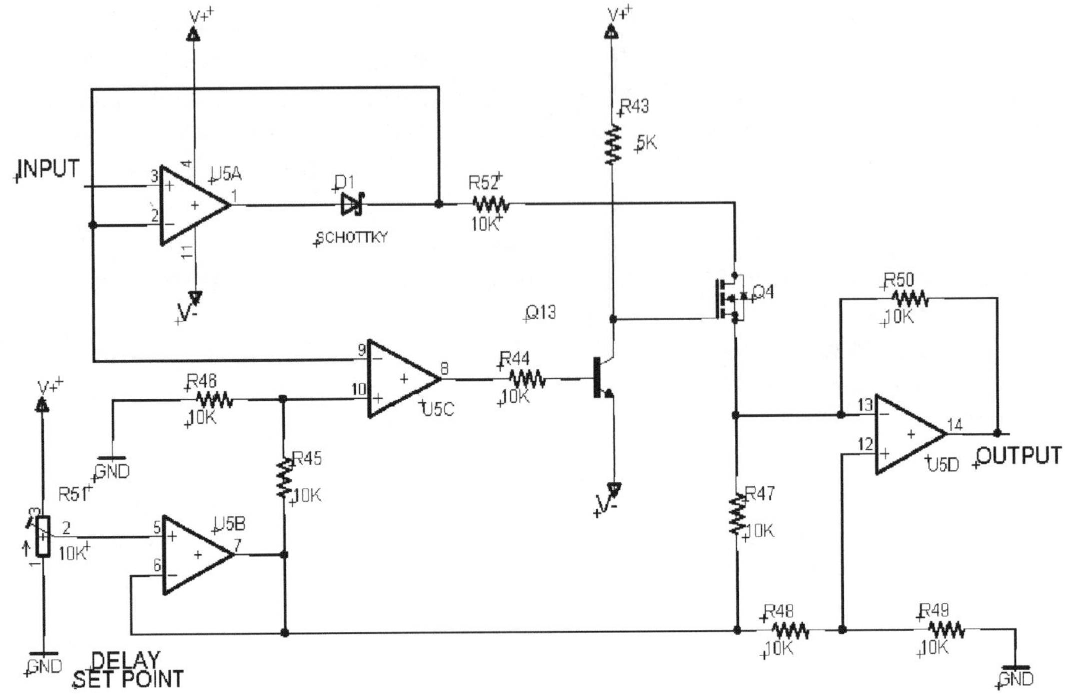

So what is this? Well, when you need a circuit that precisely outputs zero volts until some positive reference voltage has been exceeded, after which the output becomes the difference between the positive input voltage and the positive reference voltage, then this circuit may come in handy. Let's face it! There are many ways to remove the fur from a feline, but each one has its own advantages and disadvantages. The circuitry can be reconfigured for negative signals as well.

U5A is a buffer for a signal voltage. U5B is also a buffer but is used as a voltage follower for a voltage that is set by R51. U5C has, as one input the output of U1B through D1 and as the other input, half the voltage output of U5B (R45 and R46 form a divide by two resistor network). U5C acts as a comparator. Whenever the input voltage to the circuit exceeds 1/2 of the voltage set by R7, the output of U1C will become positive. This will turn off Q13 and allow Q4 to conduct. Notice that Q4's maximum gate-to-source voltage rating must be better than the difference between V+ and V-.

The voltage at the "+" input of U5D is half the value of the reference voltage (the voltage buffered by U5A) because of the resistor divider formed by R48 and R49.

The voltage at the "-" input of U5D tries to be the same as the voltage on the "+" input. In order to do this, the current flowing from the output of U5D, through R50 and R47 to the reference voltage (output of U5A) must result in a voltage at the "-" input of U5D that is half the value of the reference voltage. It turns out that the voltage at the output of U5D that results in this being true, is zero volts. Great! Now we have zero volts whenever Q4 is turned off, no matter what the reference voltage value is.

Let's say that the input signal is now more positive than the reference voltage. This causes the output of U5C to be negative. When the output of U5C is negative, Q13 turns off and Q4 fully turns on. When Q4 is in full conduction, it acts like a short. So what do we have then? Current from the output of U5B passes through D1, R52 and R47. R52 and R47 form a voltage divider, the center of which goes to the "-" input of U5D. So what is the voltage at the "-" input of U5D when Q4 turns on? The output of U5B is at or above ½ the reference voltage when Q4 turns on, no current will pass through R52 (both ends of it are at the same potential, namely ½ the reference). All of the current out of U1D will pass through R52 and R47 only. The voltage out of U5D that causes the "-" input of U5D to be ½ the reference voltage is zero volts. Great!

So the main advantage of using this circuit to maintain zero volts out of it, up to some reference level, is that Q4 is turned on when it doesn't need to conduct any current. There will be no "glitches" or sudden changes as a result. As the input voltage continues to increase, the current contribution of U5A's output (at the cathode of D1) through R52 will cause the output voltage of U5D to increase proportionately (even though it is negatively). Let's take the situation where the output of U5B is the same as the reference voltage (the output of U5A). Remember that the output of U5D will try desperately hard to maintain ½ the reference voltage on it's "-" input. Therefore, we can consider that this voltage is always ½ the reference voltage. So now we have a network of three resistors and three voltages. Assume that the voltages at the top of the resistors is 4 Volts.

Here is a diagram of the network:

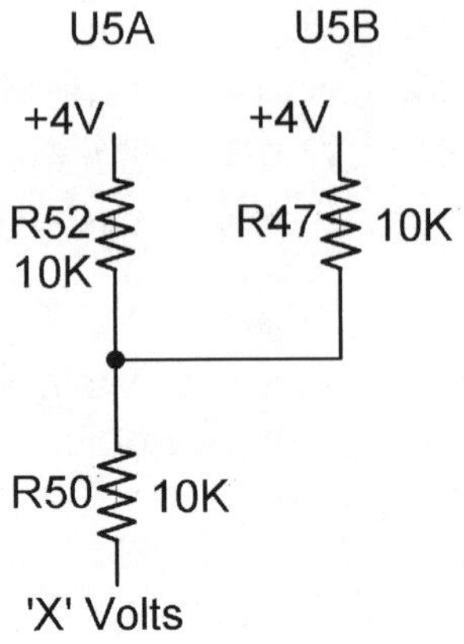

The "X" volts is the output of U1D that is needed to get 2 Volts at the junction point. The current through each of the two top resistors will be;

$$\frac{4V - 2V}{10KOhms} \quad \text{or} \quad 0.2 \text{ milliamps}$$

The total current through both top resistors will then be 0.4 milliamps. The 0.4 mA flows through the bottom resistor and causes a voltage drop of:

0.0004 x 10,000 Ohms or 4 Volts

With +2Volts at the junction point and 4 Volts dropped across the lower resistor, the value of "X" Volts must be:

2V - 4V or -2 Volts

This is the output voltage of U1D and is the difference between the reference voltage and the input voltage (inverted, of course).

Putting an inverting stage after the output of U1D makes everything positive again.

Nuff said.

Your Own Business

Believe me, there's nothing like your own business. If you want some freedom in this life, other than being independently wealthy, that's the way to go. The truth is that you'll probably put in a lot more hours than if you were working for someone else, but you won't even notice it. Oh sure, there is some drudge work. There's drudge work in anything. But you get to pick and choose where and when. That's the ticket.

So what's holding you back? Well, if you have a family you have a lot to consider. The income is initially going to be less. A lot less! How about medical insurance for everyone in the family? You'd better have a working wife with regular benefits or you're going to have to take a lot of very big chances. I had my own business when I was single. Big difference. I totally enjoyed the freedom because I didn't have that much responsibility.

As a personal note, I always felt that in a family, one of the parents had to be available for the kid(s) at all times. For that reason, my wife either didn't work at all or had a job within very close proximity to my son. Usually her jobs were part-time and didn't have good benefits. As a result I didn't feel comfortable with going into my own business. I needed the security. That's probably a common situation with engineers today.

Some engineers try to get the same feeling that they would get if they were in their own business by doing contract work. Believe me, it's not the same. 8 to 5 jobs are no way to be in your own business. Where's the freedom?

Maybe your own business involves new product development. Here are a few guidelines when doing your developing:

1. Developing a product that can be sold by mail order (UPS, FedEx...) is a big advantage. It means that you don't have to visit the customer sites in order to repair the product. They just mail it back and, if you have problems fixing it, you replace it for them.
2. Because of the internet, the product doesn't have to be a universal product but can be targeted to specific users. The internet allows them to find you so GET A WEB PAGE and a WEBSITE!
3. When you have a website, it's a good idea to put as much information about the device as possible on the webpage (however, don't give away the farm. No schematics or lay-outs). Include the capabilities of the device.

4. On the webpage, include the specifications of the device. Include any special instructions such as power supply requirements and mounting instructions.
5. Whenever possible use components that can be ordered and received within a couple of days. Having P.C. boards fabricated and keeping only completed units as stock items will help to keep component stock low.
6. Small, uncomplicated products mean that you don't need much space for manufacturing. Also, having as much assembly as possible done out-of-house means less required space.
7. With standard products and short lead-times, no large investment is required.
8. Simple circuits and simple components mean less requirement for special tools.

You may have to give away a few units if a customer accidentally hooks it up backwards, or to the wrong power source, or they accidentally flatten the unit, but that's part of starting the business. They'll remember you took care of them (even though they did a stupid thing).

In the end, all you're left with is response to customer special requests. Because It isn't the standard product line, they will have to pay a great deal, especially if it involves any design work. When you get customer requests, you'll know that you're on the way.

Follow your bliss.

Detecting Zero Crossings in a Three-Phase System

This circuit is rather specific. It is used to detect the zero crossing points of each phase in a three phase voltage source. Why? Because three phase systems aren't always balanced. The relationship between phases is not always constant. So what! Well, when you want to trigger a circuit, say an SCR, Thyristor or Power FET based upon the relationship of three phases you will definitely need to know the zero crossing points of the phases. Great.

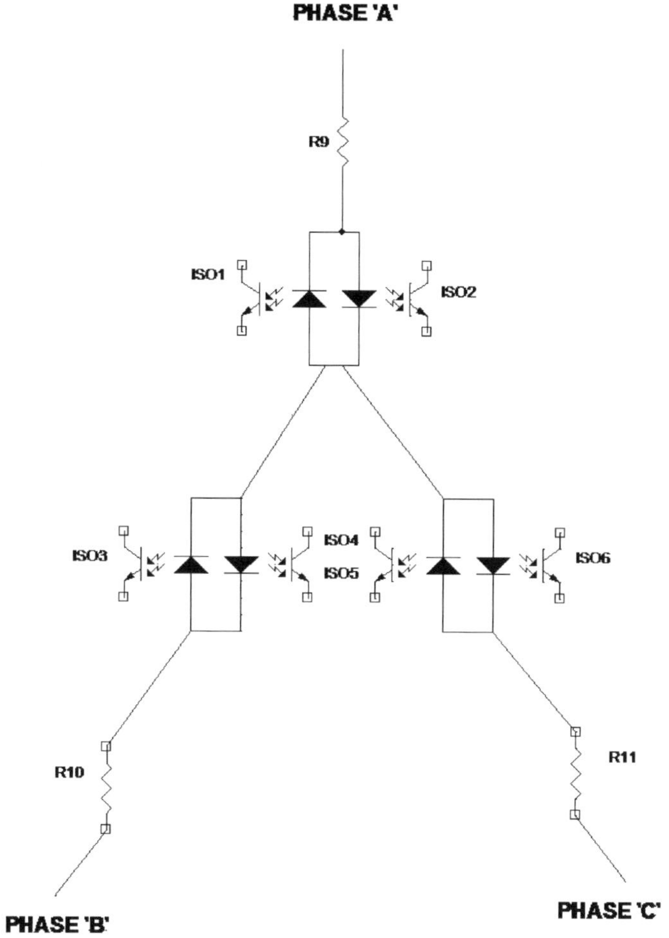

So, when the voltage at phase C is directly between the voltage at A and B, then neither of the diodes in the optos (iso5 and iso6) will be conducting. If you put the transistors associated with these diodes into an AND gate, when neither diode is conducting, the output of the AND gate will be a high level, indicating the zero crossing point for phase C (with respect to phases A and B). In most control systems, it is more important to detect the mid-point of one of the phases between the other two phases when determining the zero crossing point. Phases are often imbalanced, especially in re-generating motors with little or no load. In those cases, this detection method serves nicely.

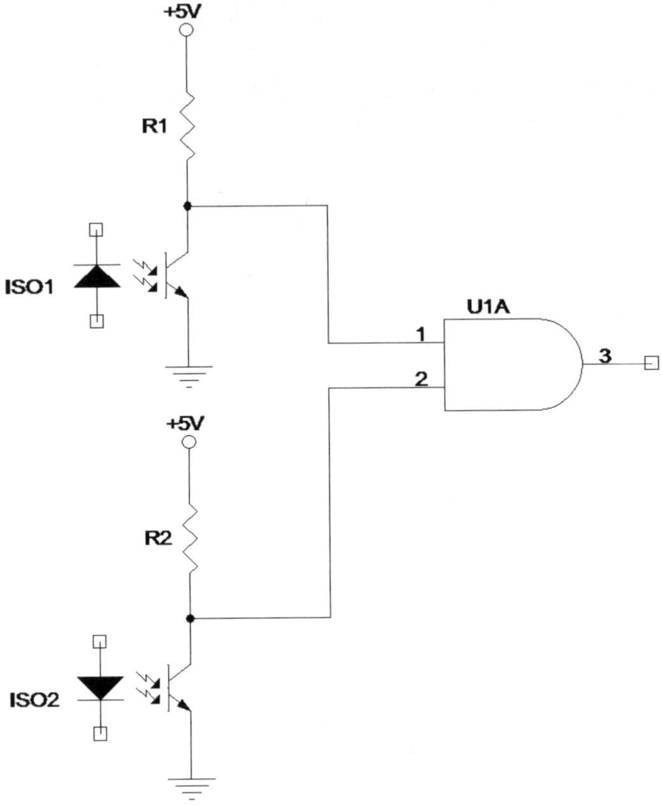

If you want to delay the triggering of an SCR, Triacs, Bipolar Transistors or FET transistors in order to limit the current through phases, this is the necessary zero crossing information for doing it.

This is the standard waveform for three phase Alternating Current. Notice that the zero crossing points of phase ‘A’ have a vertical line through them.

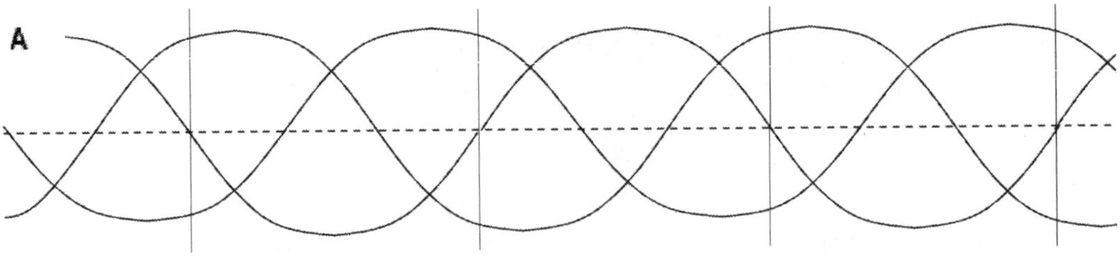

In the figure below, the areas between phases 'B' and 'C' have been darkened so that it is easier to see the point at which 'A' is precisely between 'B' and 'C'.

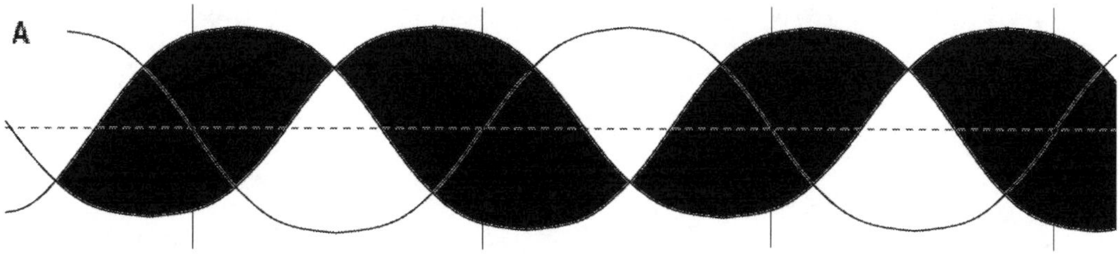

So if you're into 3-phase detection and drive circuits, this may help.

As usual, best of luck.

Dealing with Head Hunters

First, I have to immediately apologize to those placement agents that find the term "Head Hunter" offensive. Placement agents (head hunters) serve a very useful purpose. I myself have gotten at least three jobs through a placement agency. The only thing you have to keep in mind is that the placement agent works both ends against the middle. No matter what they tell you, they are only after one thing namely, the commission.

I've had many interviews that were arranged by placement agents. Most of them didn't work out. The reason for this is that a placement agent will send you in on an interview, along with a lot of other prospective clients, without fully knowing your capabilities. Let's face it, in most cases they don't know what you do. They place all types of engineers and maybe some who aren't engineers. But the fault isn't always with them. After you've been contacted, the first thing they will do is read you the requirements for the job they think may be 'right for you'. Trouble is, you are the one who says 'that sounds like something I can do'. But really, it may be disinteresting to you and you just want the job or it sounds interesting but you really don't have the background. Anyway, they have the interest of their commission at heart.

The thing to do is to take advantage of the situation as much as you can. Tell the placement agent everything that you feel is an asset. Make it clear to the placement agent what you want. They'll probably want to know what salary range you want. Just be honest. This has nothing to do what will actually happen at an interview. Make sure they understand your travel requirements and let them know that you would prefer a phone interview before visiting prospective employers. This prevents you from wasting both your time and the employer's time.

Now, I have been misled a few times. One time in particular a 'head hunter' made a Saturday morning interview for me with a prospective employer over 50 miles away and missed the interview date by one week. I got there early enough, but no-one was there. At about 8:00 A.M. someone pulled up and told me the situation. What a waste of time. Actually, I never got that interview because something came up in the mean time. Needless to say, I never dealt with that agent again.

I have always had regard for the work that placement agents do. Hell, I was in sales one summer selling door-to-door books. But I never forget that they are in it for the commission. They often have to send in several clients for the same job. You may wind up competing with these clients. It's just part of the game. The whole idea is to have a good working relationship with the placement agent. As I said before, try as best you can to define what you are looking for in order to prevent everyone involved from wasting each other's time.

Most of your communication with a placement agent will be over the phone. You'll probably never even see the guy who got you the interviews. You'll probably only meet him if you are hired by one of the companies that he consistently represents, because he will visit them in order to maintain his client. Also, placement agents work together, like real estate agents. They take part of a commission when they share the client. So you may get calls from other placement agents once you have made contact with one agent or agency. Don't let this bother you, just make sure that they understand what your experience is and what you want from the job. If your don't, if you're desperate you'll probably wind up with some job. Not the right job, not a job that you'll stay with, just a job.

O.K., O.K, at least it's a job. When you have responsibilities you'll do just about anything to put food on the table. There is no better motivation for self improvement than the necessity of having a job. Believe me, I know! I had good jobs for several years. Then things took a down turn. I wound up getting laid off and having to take a crummy job, actually two crummy jobs one after the other. This motivated me to finish my engineering undergraduate degree and then to get my Professional Engineering registration and a Masters degree in electrical engineering. With that under my belt I could at least get jobs that weren't as crummy. "You gotta take what they're givin' when you're wokin' for a livin.'"

Good luck on the hunt.

Binary Division and Multiplication

These are things that I have come back to time and time again. I usually work with 8-bit processors or micro-controllers but the flow should work with any bit machine. The thing about electronics, circuit design and software scripts is that you keep returning to them. The problem is, you usually don't return often enough to recall exactly how you did it last time. Well, here is one way of dividing. The main elements are the dividend, the divisor and the quotient. A flow diagram follows:

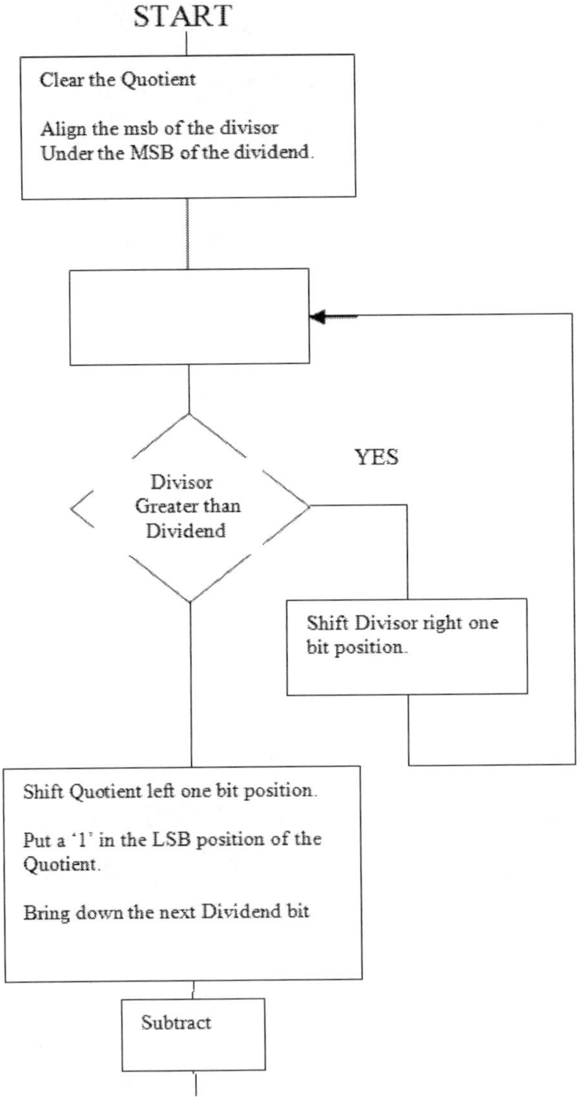

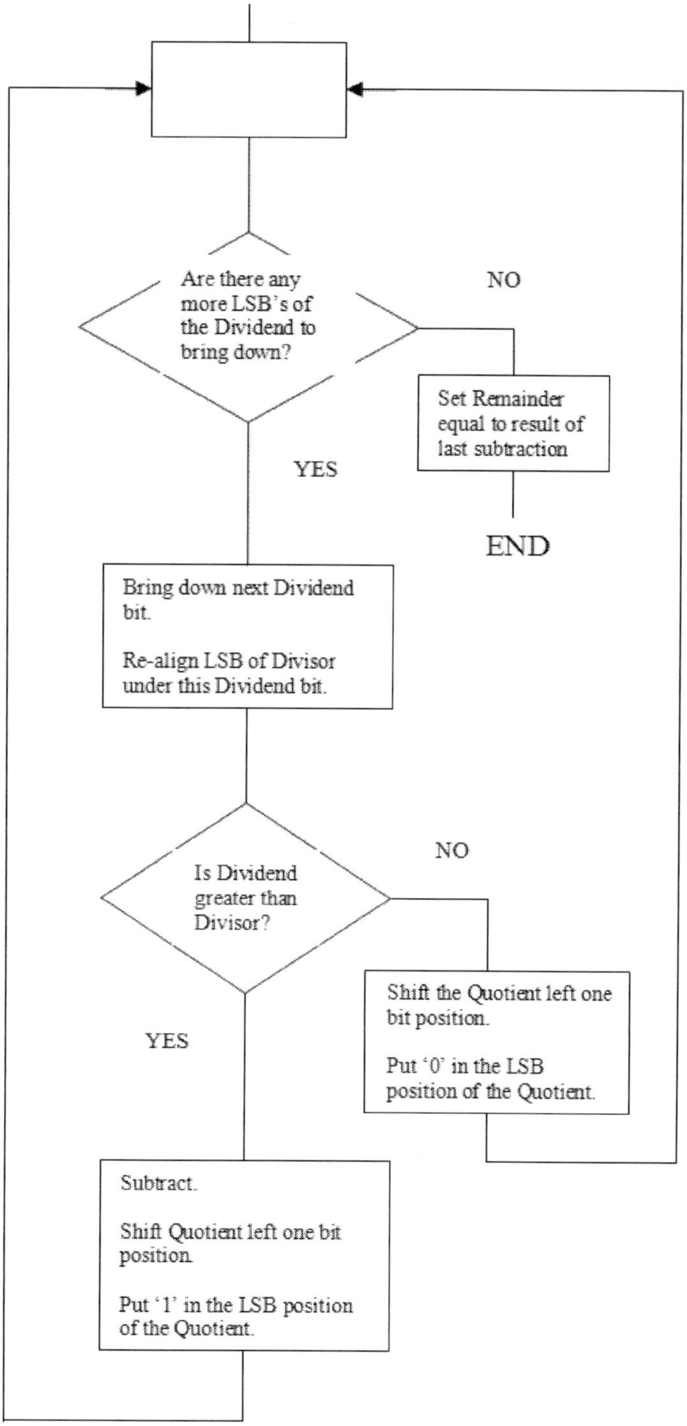
Are there any more LSB's of the Dividend to bring down?
NO
Set Remainder equal to result of last subtraction
END
YES
Bring down next Dividend bit.
Re-align LSB of Divisor under this Dividend bit.
Is Dividend greater than Divisor?
NO
Shift the Quotient left one bit position.
Put '0' in the LSB position of the Quotient.
YES
Subtract.
Shift Quotient left one bit position.
Put '1' in the LSB position of the Quotient.

So let's take an example. First step is to clear the quotient ('0000000'). Make the dividend equal to '1000101' (69 decimal) and the divisor equal to '101' (5 decimal). Start at the beginning of the flow diagram.

Aligning the MSBs of the dividend and divisor yields;

1000101
101

For the alignment shown, the divisor is bigger than the dividend (101 is larger than 100). So, according to the flow diagram, we shift the divisor one place to the right and get;

1000101
 101

So now the divisor is greater than the dividend ('1000' is larger than '101') so we subtract the two. The result is '011' (8 – 5 = 3) so we shift the quotient one bit position to the left ('00000000') and put a '1' in the LSB position. Therefore the quotient equals;

0000000**1**

Now bring down the next bit of the Dividend and place it to the right of the LSB position of the results of the last subtraction ('011').

011**1**
 101

The result is larger than the divisor so shift the quotient one position to the left ('000000010) and put a '1' in the LSB position ('000000011').

So the result of the subtraction is '010' (7 – 5 = 2).
There are still more LSBs of the dividend left so bring down the next LSB

010**0**
 101

Is this dividend larger than the divisor? No! So we shift the quotient one position to the left and put a '0' in the LSB position ('00000000110')
There is one more LSB of the dividend to bring down so bring it down and re-locate the LSB of the dividend with this new bit.

0100**1**

101

The result ('1001') is larger than the dividend ('101') so shift the quotient one position to the left and put a '1' in the LSB position ('000000001101').

Subtracting these leaves '0100' (4). This is smaller than the dividend and there are no more LSBs of the quotient to bring down. So the **remainder is '0100'**, or 4 decimal.
Summing up, we divided '1000101' (69 decimal) by '101' (5 decimal) and got '000000001101' (eliminating the leading zeros, 13 decimal) and a remainder of '0100' or 4 decimal. This checks,
that's it!

Binary Multiplication example:

The whole operation is a 'shift and add'. Great! But when? Let's take an example.

Given:

Multiplicand	=	110101	=53
Multiplier	=	11011	= 27
Resultant	=	?	= 1431

First, set the Resultant equal to all zeros (00000000).

Then, ask the question 'does the LSB of the Multiplier equal '1'? In this case it does. So the Resultant takes on the value of the Multiplicand (110101). If this were not the case (LSB not equal to '1') we would just continue on to the next step, which is;

Does the next significant bit of the Multiplier equal '1'? Again, this is true. So the Multiplicand is shifted left <u>*one*</u> bit position and then added to the Resultant (1101010 + 110101 = 10011111). This is the value of the new Resultant. If this were not the case we would just continue on to the next step, which is;

Does the next significant bit of the Multiplier equal '1'? Here it does not, so we just continue on (Resultant = 10011111). If it had equaled '1', we would have shifted the Multiplicand left <u>*two*</u> bit positions and then added this to the Resultant to get a new value for the Resultant.

Does the next significant bit of the Multiplier equal '1'? Yes it does. So shift the Multiplicand left *three* bit positions and add this result to the Resultant (110101000 + 10011111 = 1001000111) to get a new value for the Resultant.

Does the next significant bit of the Multiplier equal '1'? You bet! So shift the Multiplicand left *four* bit positions and add the result to the Resultant (1101010000 + 1001000111 = 10110010111) which equals 1431 decimal.

With more bit positions this operation would continue until there were no more Multiplier terms. The final Resultant is the answer. When programming, you'll have to set up the 'DO LOOPS' or whatever, for the number of bit positions in the Multiplier.

That's about it.

Serial A/D, D/A Circuit with Selectable Bit Resolution

Now, this is a bit different from the usual A/D. The circuitry is simple but slightly more elaborate than the usual A/D or D/A. The heart of this D/A is a matched pair of resistors in a circuit that looks a lot like an instrumentation amplifier.

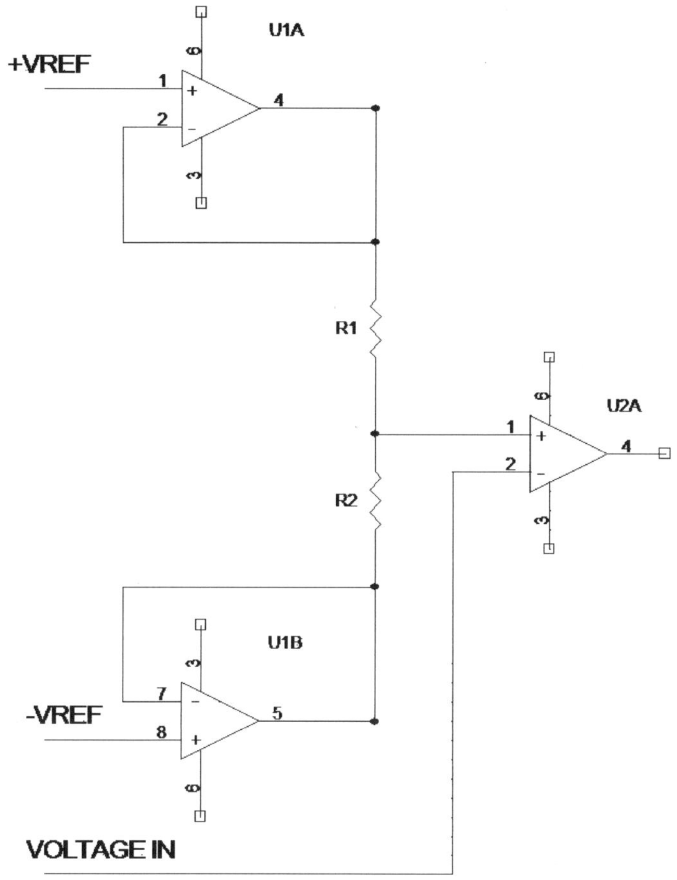

Let's see how the principle works on an A/D conversion. The first thing is to capture the analog input. This would be 'VOLTAGE IN' on the circuit above. Then the input is compared to the voltage between R1 and R2. For the sake of example let's make the signals to be converted have a range of 0 Volts to +10 Volts. So in our circuit we let '-VREF' equal 0 Volts and '+VREF' equal +10 Volts. The voltage at the junction or R1 and R2 will be +5V. The output of comparator U2A will be the result of comparing the 'VOLTAGE IN' and the voltage at the junction of R1 and R2. O.K! So what good is this? Well, not much unless we use this piece of information. We need to add a little circuitry now. Things might be better explained if they are in block diagram form. So here goes:

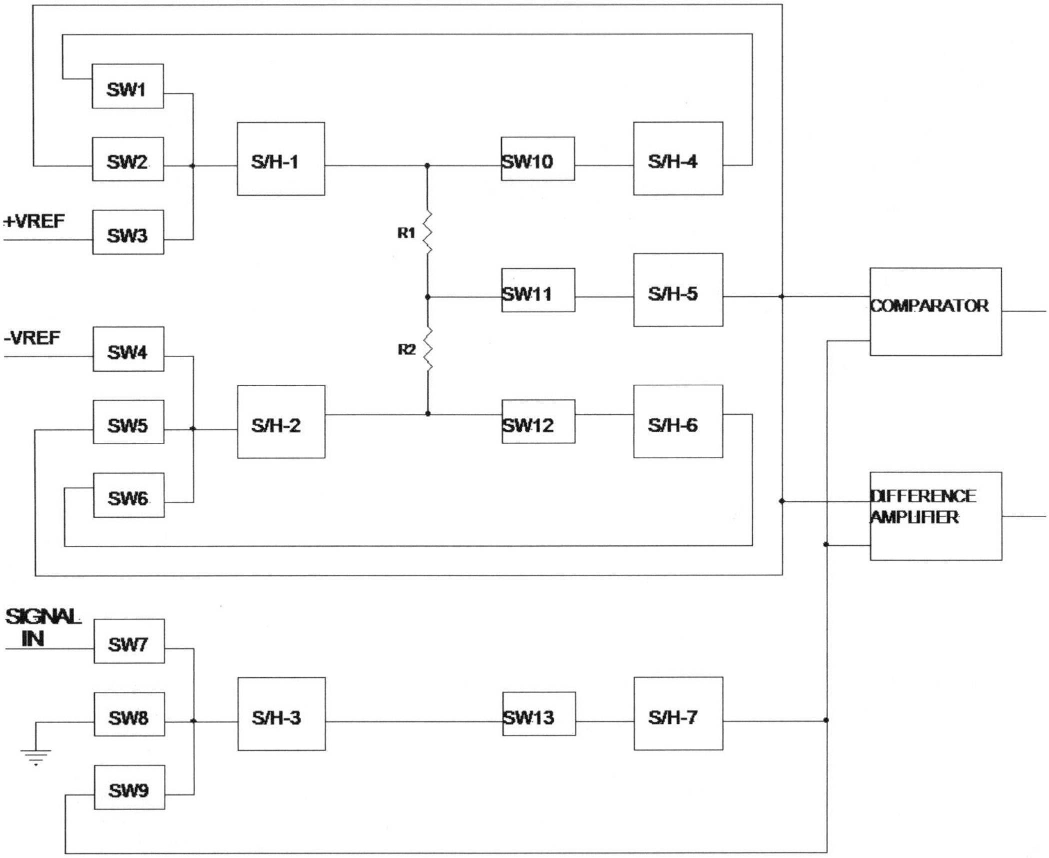

Holy Cow! Things have gotten a bit more complex in a hurry. O.K., O.K, complex but not complicated. We've added some FET switches, some Sample-and-Holds and a Difference Amplifier that, I don't believe even belongs in the A/D Conversion description. Because of all of the FET switches, some form of control circuitry is necessary. These must also be some form of clock involved. A CPLD or micro-controller or some form of state machine will be required. So how about an explanation?

To start with, Switches SW3, SW4 and SW7 are closed. All other switches are open. Sample-and-Holds S/H-1, S/H-2 and S/H-3 track these inputs. The three switches are then opened and the voltage which is half way between –VREF and +VREF appears at the junction of R1 and R2. Switches SW11 and SW13 are then closed for a brief period of time (one clock cycle maybe?) and then opened. These voltages are held in S/H-5 and S/H-7. S/H-5 holds the voltage at the junction of R1 and R2 and S/H-7 holds the SIGNAL IN voltage. These voltages are then compared in the COMPARATOR. If the voltage of S/H-5 is greater than S/H-7 the output of the COMPARATOR will be some level. For the sake of example let's make this a low level. This 'bit' now becomes the MSB of the conversion and indicates to the control circuitry that the SIGNAL IN has a voltage level that is less than half of the difference between +VREF and –VREF. Sensing this condition, the control circuitry will then cause the output of S/H-5 to be fed back to S/H-1 through SW2. This effectively forms a new '+VREF'. The SIGNAL IN will still be in the range between this new '+VREF' and –VREF. Here's a bit of a trick. In order to minimize the offset differences due to different signal paths, the control circuitry will also feed the output of S/H-6 back through SW6 to S/H-2. For the same reason the control circuitry also feeds the output of S/H-7 back to S/H-3 through SW9. All of these feed-back operations occur at the same time.

Now, let's say that the SIGNAL IN is greater than half of the difference between the new '+VREF' and –VREF. This produces a high level at the output of the COMPARATOR, which is the value of the next bit of the conversion. This is indicated to the control circuitry which causes the output of S/H-5 to be fed back to the –VREF side to form a new '-VREF'. Once again, the SIGNAL IN will be between the new '+VREF' and the new '-VREF'. This operation will continue for as long as the control signal dictates. The number of bits of resolution of the conversion will coincide directly with the number of feed-back cycles.

O.K! How about the D/A operation? This process is the reverse of the A/D. A series of low and high level bits are read by the control circuitry and cause the circuit switches to appropriately feed back signal levels in order to produce the analog representation of this series of digital bits. SW6 will input a ground level to S/H-3 which will pass through SW13 and arrive at the output of S/H-7. This forms the reference for the converted voltage. +VREF will go to +10Volts and –VREF will go to ground. The output of the DIFFERENCE AMPLIFIER will contain the resulting analog voltage.

So let's see, the control circuitry seems to have a lot to do. In addition there has to be some type of 'serial-load, parallel-out' shift register in order to make the A/D useful. Then too, the analog output signal from the difference amplifier is not held in any way. It will deteriorate with the hold time of the sample-and-holds, although this may be aided by continuous conversions.

Hardware-wise, the trick is to keep circuit elements as balanced as possible. Keeping S/H-1, S/H-2 and S/H-3 in the same package is one way. Same goes for S/H-4, S/H-5, S/H-6 and S/H-7. Admittedly, the number of bits of resolution can become ridiculous. Based upon the application, the devices used and the desired resolution, it is possible to arrive at a very high number for the bits of resolution.

How about uses for this circuit. The whole thing came about as a circuit used to pass a slowly changing analog voltage across a barrier with a difference of potential of 250,000 Volts. Two identical circuits were used. One of the circuits acted as the transmitter (the A/D) and the other circuit acted as the receiver (the D/A). The digital sequence was passed from source to receiver through a fiber-optic link. The thing to remember is that the conversions had to be synchronized. The control circuitry at the receive end had to know when to expect the first bit (MSB). An additional fiber-optic link was used to carry the conversion clock. So things are a bit more complicated. Some LED drivers and receivers had to be added. But, what the hey. It's a quarter of a million volts.

Here is a flow diagram that could be used when coming up with the algorithm for the control circuitry:

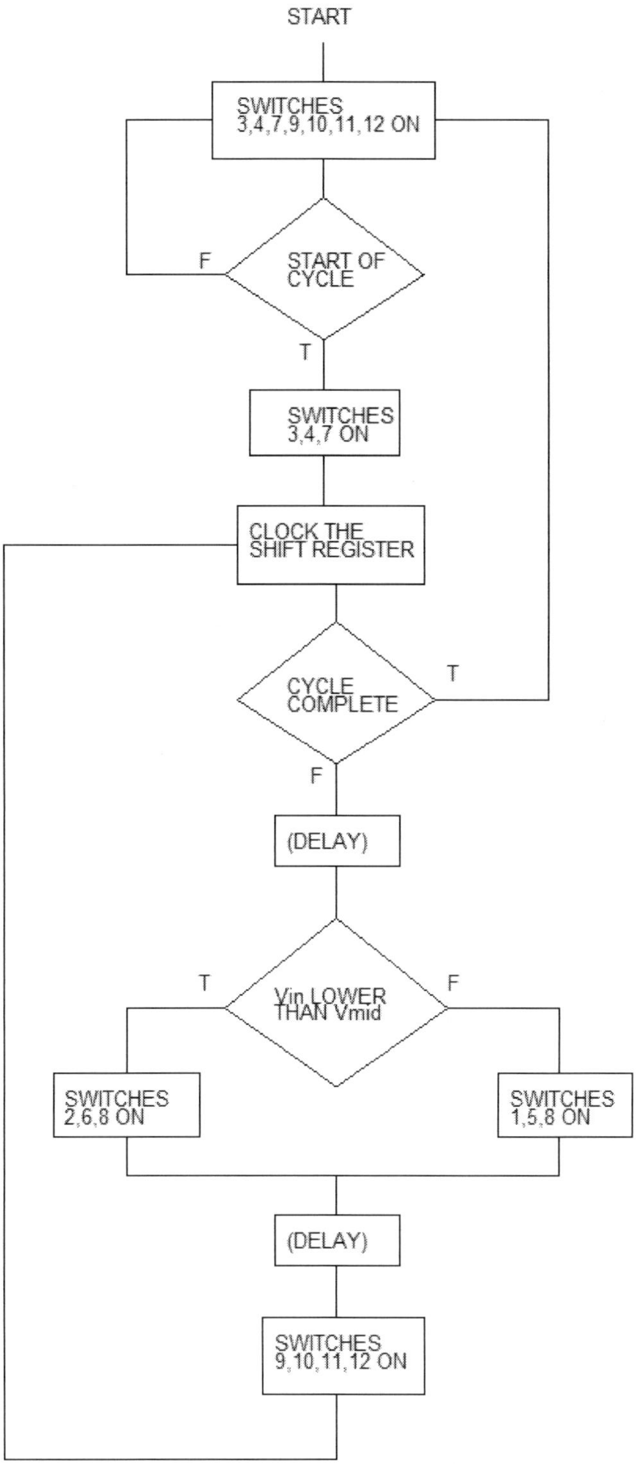
START
SWITCHES 3,4,7,9,10,11,12 ON
START OF CYCLE
F
T
SWITCHES 3,4,7 ON
CLOCK THE SHIFT REGISTER
CYCLE COMPLETE
T
F
(DELAY)
Vin LOWER THAN Vmid
T
F
SWITCHES 2,6,8 ON
SWITCHES 1,5,8 ON
(DELAY)
SWITCHES 9,10,11,12 ON

Circuitry for the A/D-D/A might look something like this:

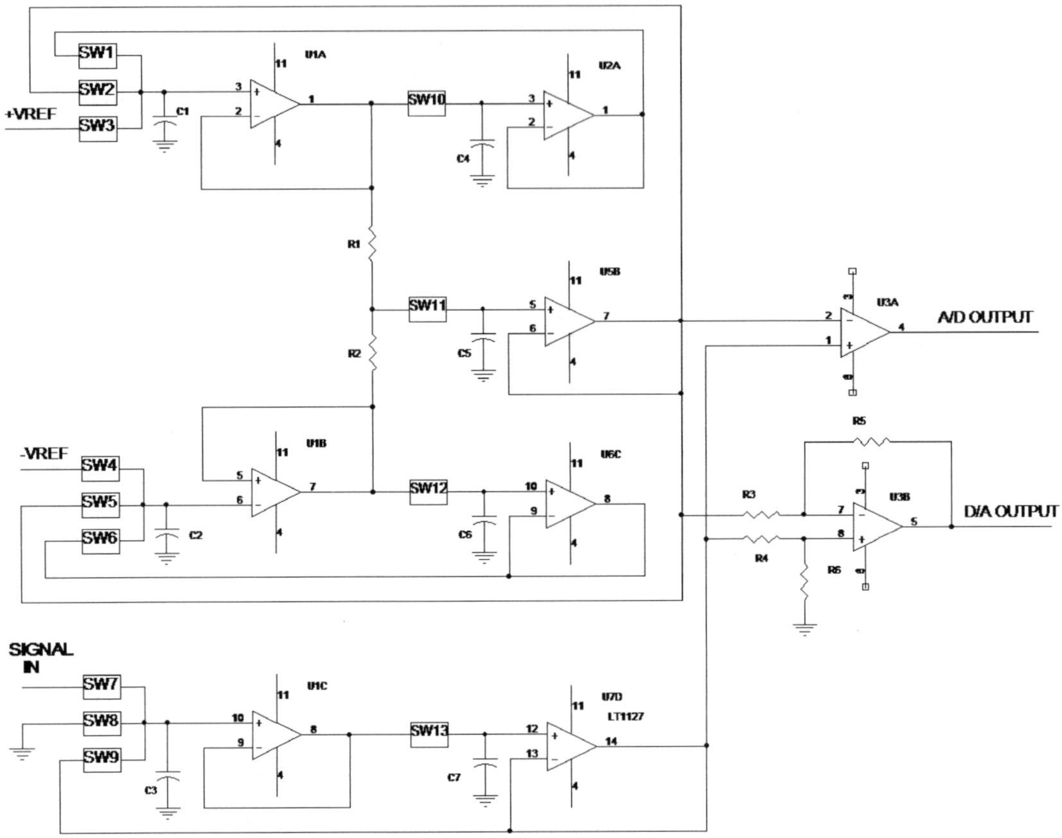

Well, that’s about it.

Voltage Doublers and Negative Voltage Circuits

Let's say you have some CMOS logic that runs on 10Volts and that you need a fairly low-current supply of between 15 and 20Volts DC. Well, the circuitry below may fit the bill. Remember the oscillator circuit we made from a Schmitt Trigger Inverter? We're going to need an oscillator in this circuitry also.

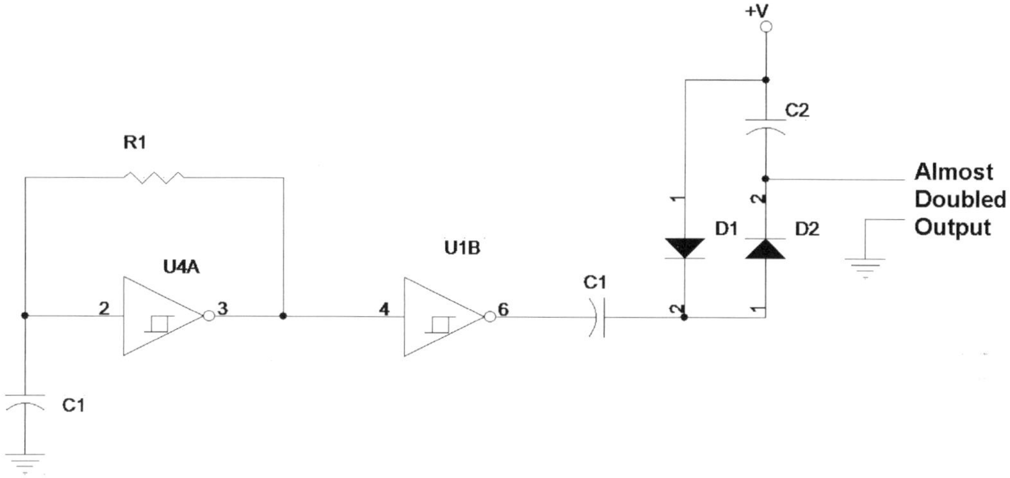

U4A, R1 and C1 make up the oscillator. You can choose any frequency you like. Something around 25KHz sounds good. Anyway, U1B is a buffer stage. Inverting the signal has nothing to do with it. The output of U1B will see a short to +V during operation but this is taken care of by the current limit within U1B.

How do we get nearly twice the voltage at the cathode of D2? When the output of U1B is low, C1 charges to a little less than +V through D1. When the output of U1B switches to a high it references the left side of C1 to +V and, since C1 is charged to +V, the voltage at the right side of C1 is almost twice +V. The cathode of D2 sees this 2 *+V and charges C2 to 2 *+V (minus the diode drops of D1 and D2). If the load (at '+2V') doesn't require a great deal of current, this voltage will come close to 2 *+V. In order to get the highest voltage you may want to use Schottky diodes for D1 and D2. These have approximately 0.2V drop as compared to the 0.7V drop for PN junction diodes.

What about voltages below Ground? Take a look at the circuit below.

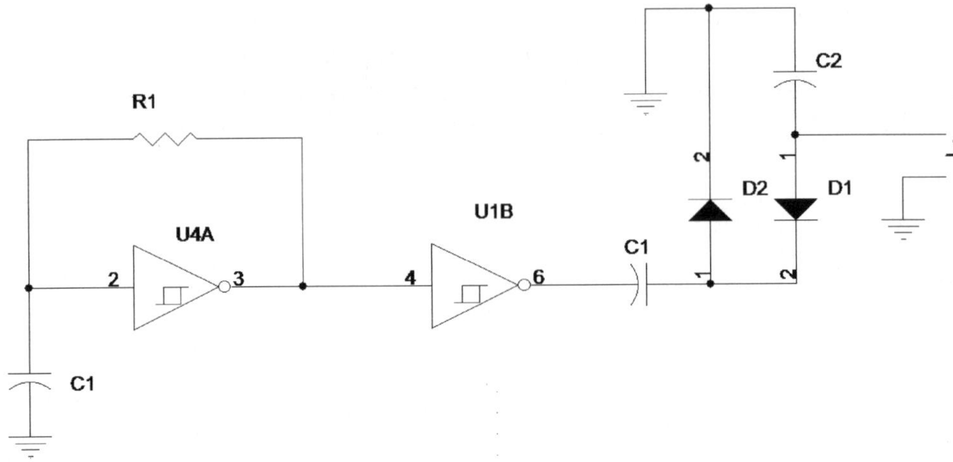

Looks about the same, except for the diode directions and the connection to ground instead of +V. O.K.! When the output of U1B is low, nothing much happens. When U1B's output goes high, C1 charges, through D2, to this +V level. NOW when U1B goes low the left side of C1 is at near ground potential and the right side of C1 is at –V. D1 is now forward biased and charged C2 to this –V potential.

So we now have a circuit for generating almost +20V and one for generating –V. With 10 Volts for +V, we can get fairly close to a +28V difference. One more thing. There is no reason why the outputs of CMOS logic can't be hooked together. So we can parallel several devices and wind up with a much higher current capability for these supplies.

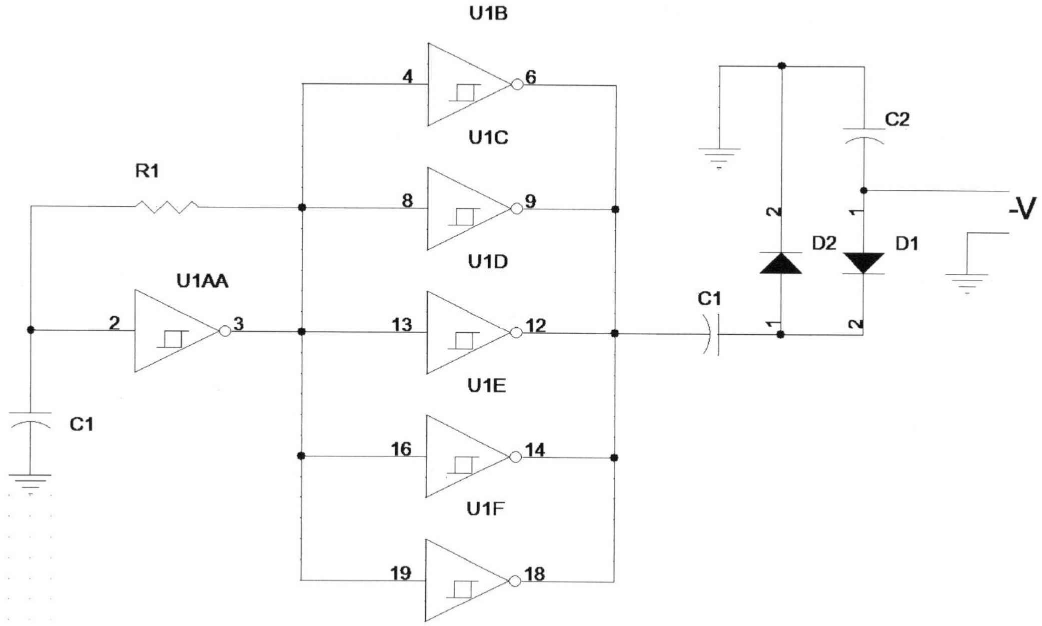
U1B
4
6
U1C
C2
R1
8
9
-V
2
1
D2
D1
U1D
U1AA
C1
2
3
13
12
1
2
U1E
C1
16
14
U1F
19
18

Good luck.

Strobe Firing Circuit

This circuit was originally intended for firing a strobe lamp. A trigger transformer is used to trigger the strobe lamp. Trigger transformers can be bought in several configurations. For the sake of example, the voltage across the secondary winding of this transformer is approximately 20,000 Volts when a voltage of 200volts is impressed across the primary winding.

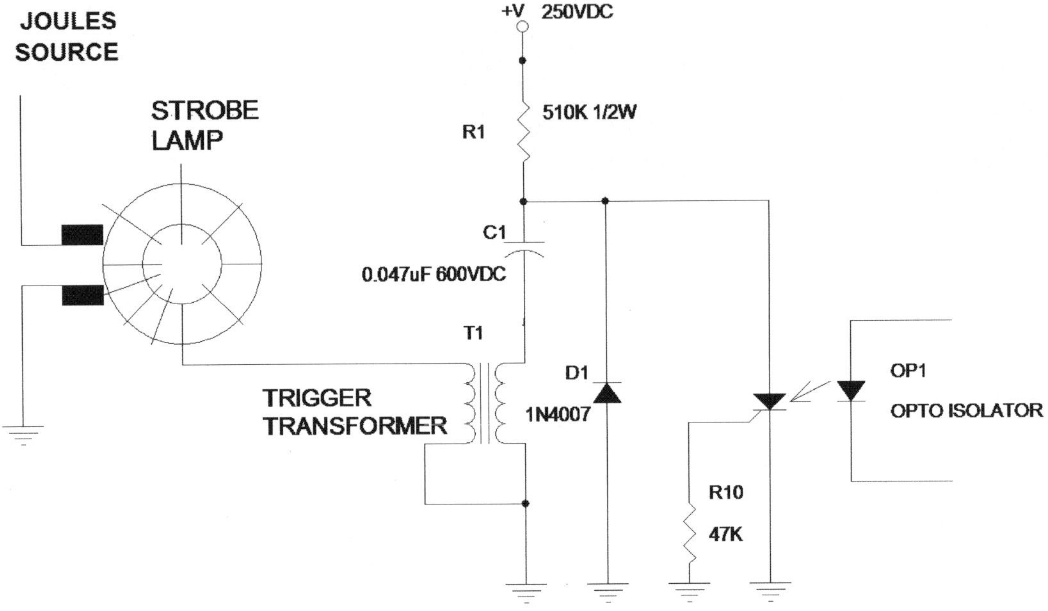

A little about the strobe lamp:
The strobe lamp itself is made of glass or quartz and is filled with an inert gas or mixture of inert gases such as argon, xenon or neon. Through the use of different mixtures, the voltage required for ionizing the gas in the tube can be made lower or higher. Also by varying the mixture the life of the tube can be varied as well as the brightness of the flash. So much for the lamp. How do we trigger the darned thing?

About the circuit:
First, notice that the secondary of the trigger transformer has one side connected to ground and the other side connected to nothing except a few turns of wire wrapped around the outside of the strobe tube. The high potential on this secondary is merely felt through the tube, through the gases and at the ground connection of the strobe lamp itself. This effectively makes the secondary of the transformer a complete open circuit until the strobe lamp 'fires' or breaks down

and the gases inside ionize. Even then, only a miniscule amount of current will pass from this wire to ground. Because of this, the polarity of the transformer secondary voltage doesn't mean anything. If it is -20,000 Volts below ground or 20,000 volts above ground potential the results will be the same, ionization of the gases. So what happens before the opto-SCR turns on? How does the firing circuit get started?

The current through the R1, C1 and 'Trigger Transformer' circuit is seriously limited by R1. The transformer coil resistance and/or inductance are not even a factor. C1 will charge slowly toward 250VDC through R1. How long will it take? Let's say that when C1 is charged to 200Volts, that's plenty of coulombs, enough to trigger our strobe tube. So the charge time, from 0V across C1, is:

200/250 = 1 - e^(-t/[510KOhms * 0.047uF])

0.8 - 1 = - e^(-t/0.02397)

-0.2 = - e^(-t/0.02397)

0.2 = e^(-t/0.02397)

ln (0.2) = - t/0.02397

-1.6094 = - t/0.02397

0.0386 = t

Not bad. So the maximum rate at which the strobe firing circuit can be triggered using these values is about 26 times a second (1/0.0386), that is, if the C1 totally discharges every time the trigger circuit is fired. If C1 doesn't totally discharge the triggering rate will be higher.

O.K. Now C1 is charged and we turn on the opto-LED. The opto-SCR will conduct. Let's assume that it looks just like a short circuit. The circuit then becomes as follows:

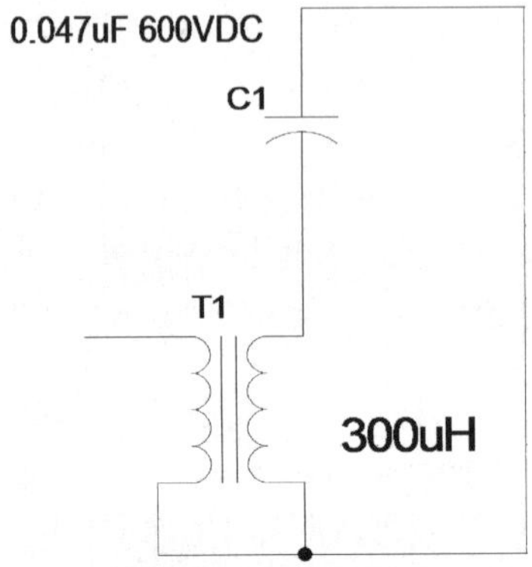

Remember that C1 is fully charged (positive potential on the top of C1). The top of C1 will suddenly be at ground potential. This makes the potential at the bottom of C1 -200VDC with respect to ground. So what! We need some current to pass through T1's primary in order to induce a voltage into the secondary of T1. The SCR is forward biased and C1 starts discharging through the SCR and L1. Now we're getting somewhere. As the current through T1's primary increases, it builds up an electro-magnetic field about T1. Remember that the load on the secondary of T1 is almost nil (just an ionizing potential is required, no major current flow). So a substantial amount of electro-magnetic energy is stored in T1.

As the current produced by C1 goes to zero, The rate-of-change of current in T1 also goes to zero. This allows T1's electro-magnetic field to begin collapsing. The collapsing field produces a current in T1 that has the same direction of flow as the current that was produced by the discharge of C1.

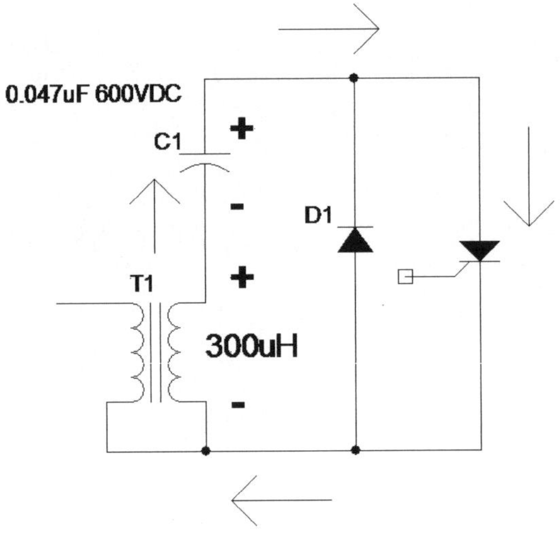

So C1 starts charging again, but now in the opposite direction (negative potential on top of C1) in the primary of T1. Eventually the field of T1 will diminish leaving this reverse charge on C1. Guess what! C1 will start discharging. A field will build about T1. When C1's current stops changing, T1's field will collapse producing a current in the same direction as the one that was produced by C1 and, happy happy, C1 will have a positive potential on the top once again.

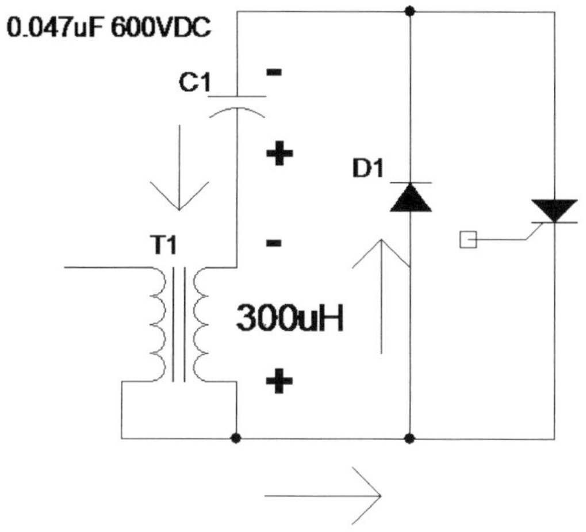

Wait a minute! This current, which is in the opposite direction, causes the SCR to be reverse biased and no current will flow through it. That's the reason for D1. D1 will carry this reverse current and at the same time allow a small reverse bias on the SCR. This reverse bias is sufficient to turn off ("sweep out" current from) the SCR. So at the end of this triggering cycle, we are left with a positive charge on C1. The time needed to charge C1 in preparation for the next trigger pulse will be greatly reduced.

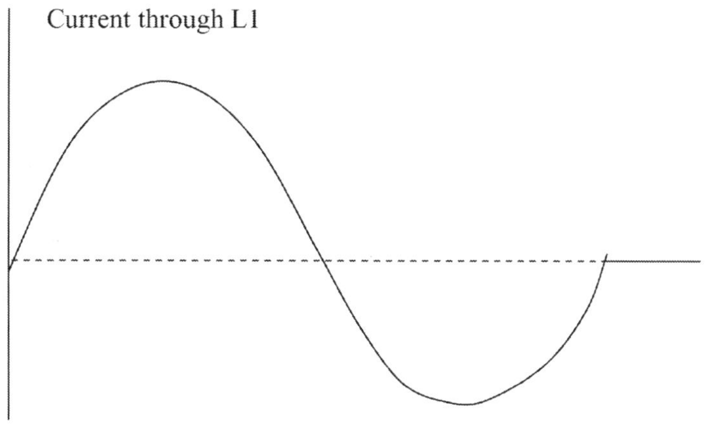

The SCR will stay 'OFF' once forward biased as long as the 'gate turn-on' signal is not present. Therefore, 'gate turn-on' signals have to be shorter than one cycle of the LC combination.

What is the duration of one cycle of the LC circuit anyway? Well, the best way to figure out the resonant frequency of the LC circuit is to use the formula;

Resonant Frequency = 1 / 2*pi*(LC^ -2)

or, 'one over the product of 2*pi and the square root of L*C'

The duration of one cycle is 2.359 x 10^ -5 seconds. This is approximately 42,380KHz, much faster than the charge time of C1 through R1.

How about the resistor from the gate of the SCR to ground? Its whole purpose is noise immunity. We can't have the gate of an SCR swinging in the breeze.

That's about it. Be mindful that this circuit just ionizes the gas in a strobe tube. We need another couple of Joules of energy to pass through the tube in order to get a flash of light out of it.

But that's another story.

Two-Phase-Plus-Neutral Motor Control Device

The most common industrial electrical power source is 3-phase power. This consists of three varying potential differences. These potential differences are in both positive and negative directions, that is, current will flow in both directions in the circuit attached to any of the three phases. The relationship of one phase to the two other phases is in degrees of shift between them. Any phase has a 120 degree phase difference with each of the other two phases.

Conventional 3-phase A.C. signal:

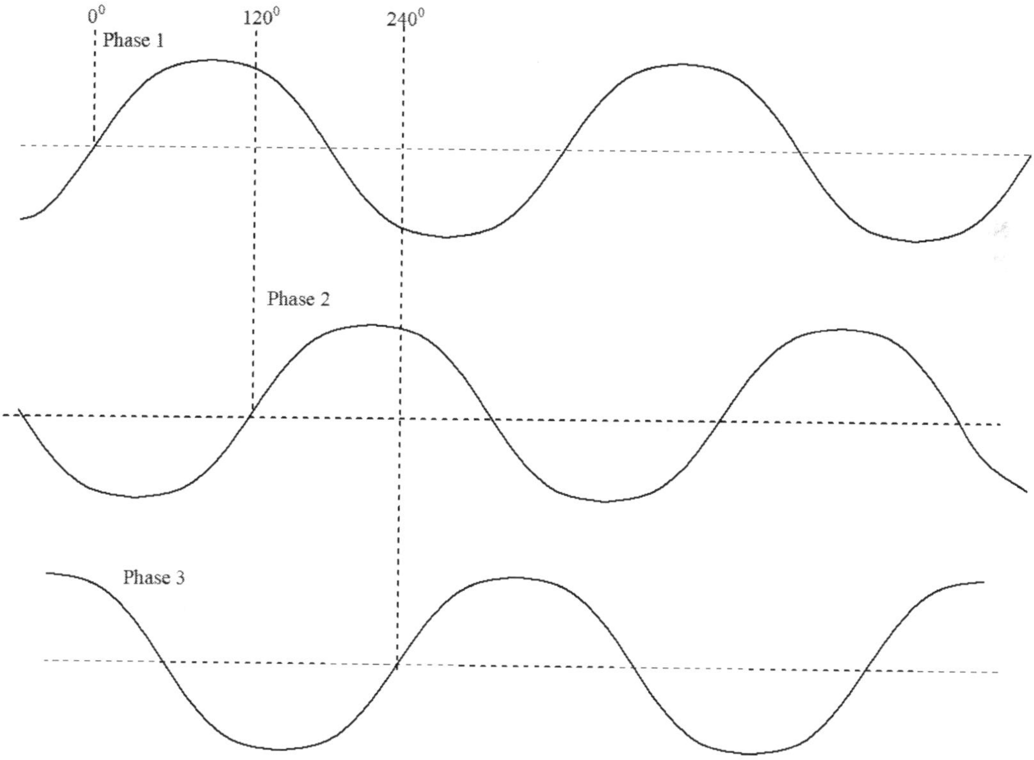

When a motor is connected to a three-phase alternating current source the voltage waveforms on each phase (with respect to neutral) are as shown in the figure below. The current in each phase for one full cycle, is also shown.

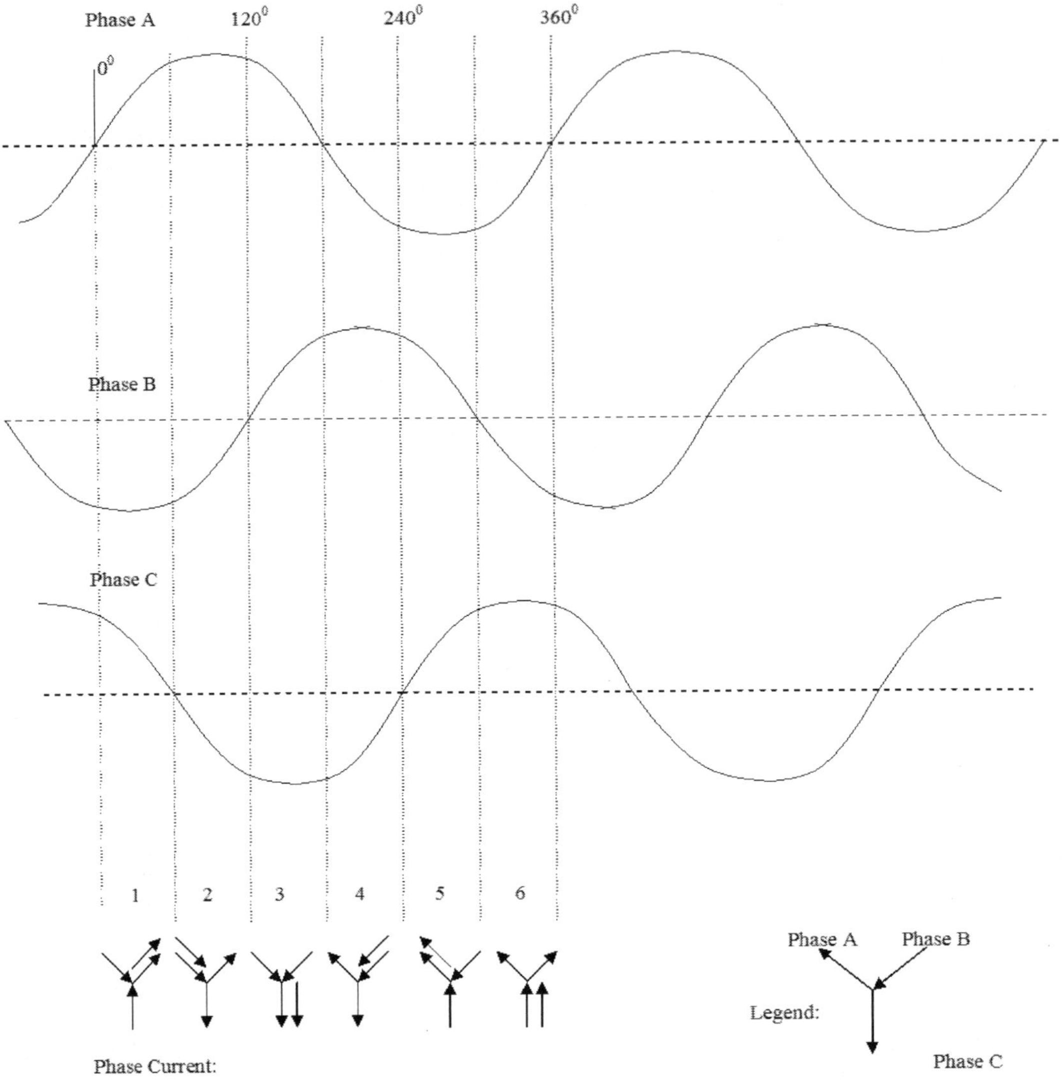

Three phase motor controllers: Solid state motor controls first used SCRs (Silicon Controlled Rectifiers) that could switch on and switch off large currents at relatively high voltages. These controllers produced what is commonly known as a "six-step" waveform. The waveforms of the three output phases of a typical six-step, three-phase motor controller are illustrated in the figure below. More recent motor controllers use fast, high voltage blocking high current switching transistors (IBGTs or FETs) and are capable of producing higher frequency, pulse width modulated output signals. These are called Pulse Width Modulated, or PWM controllers. Three-phase solid state motor controllers create a variable voltage D.C. bus from the incoming three-phase A.C. line voltage. This D.C. bus voltage is then turned back into a variable frequency three phase power source for the motor. The two-phase-plus-neutral controller works in the same way except that both a positive and negative D.C. bus voltage (with respect to the neutral of the three-phase-plus-neutral) must be created. Explanation of the conventional three-phase controller : The conventional three phase power system consists of three sine waves of voltage that are 120 degrees out of phase with each other. When applied to a balanced resistive or inductive load in either a Delta or STAR (" Y ") connected configuration, a balanced system is formed. In the following descriptions, the " Y " or STAR connected load will be used. The additional center point in a " Y " connected system is more convenient when describing current flow into and out of the three phases. The current at the star point of the Y connected configuration is constant. This is because, at any time ' t ' , the current entering the star point equals the current leaving the star point. The sum of the currents entering and leaving is zero.

Three-Phase switched controller signal with difference between phases of 120°

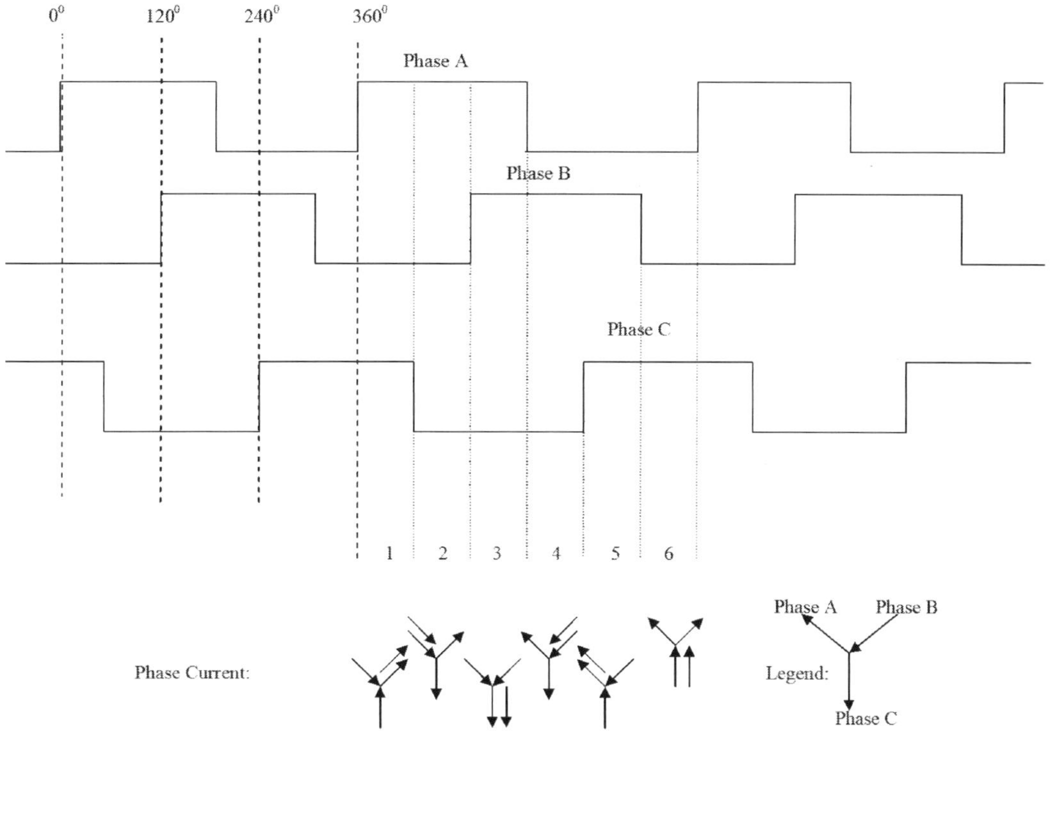

Explanation of the two-phase-plus-neutral controller:

There is another way of creating the same relative voltage differences as those of three-phase sources. If we create two Alternating Current phases that operate about a neutral voltage and have a phase relationship of 60 degrees between them, then measurements of the current in each leg of a three-phase load driven by these two phases and a neutral will be the same as the currents in a three phase load driven from a three phase source. The following figure illustrates how the two phases need to be switched in order to achieve the desired result of simulating a three-phase source. Remember that the third leg of the three-phase load is attached to Neutral and that current can travel both into and out of this neutral. Two-phase switched power with relationship between phases at 60 degrees.

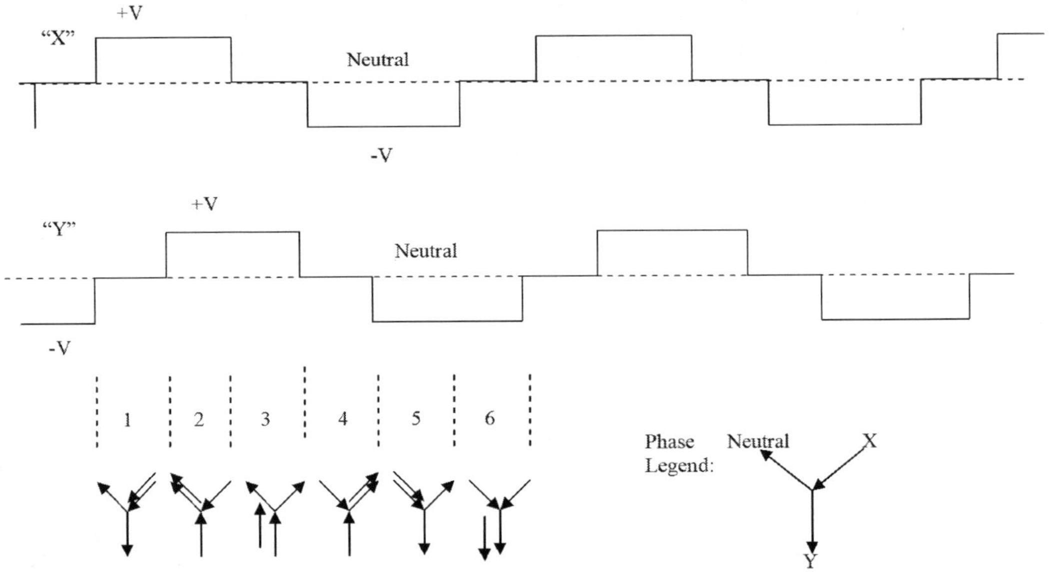

The Neutral, or dashed line in the illustration either sources or sinks current. This means that to produce this waveform across two phases there must be two distinct voltage sources; the positive voltage source must source current, the negative voltage source must sink current and the neutral (the dashed line) must allow current to flow in and out of it. This results in a complicated arrangement of drive and power circuitry. Having the neutral potential on the two phases that are 60 degrees apart means additional power components would be required. In order to overcome this requirement, no neutral voltage will be used on the two phases that are 60 degrees apart (remember that neutral is the third phase of this three phase connection). Instead, at those times when the voltage of either phase should be neutral (at the level of the dashed line) the voltage of that phase will rapidly alternate between positive and negative. The result of the average will be a neutral voltage on that phase (see figure 7, below). Notice that the average current in each phase is the same as that of the previous two phase and three phase sources.

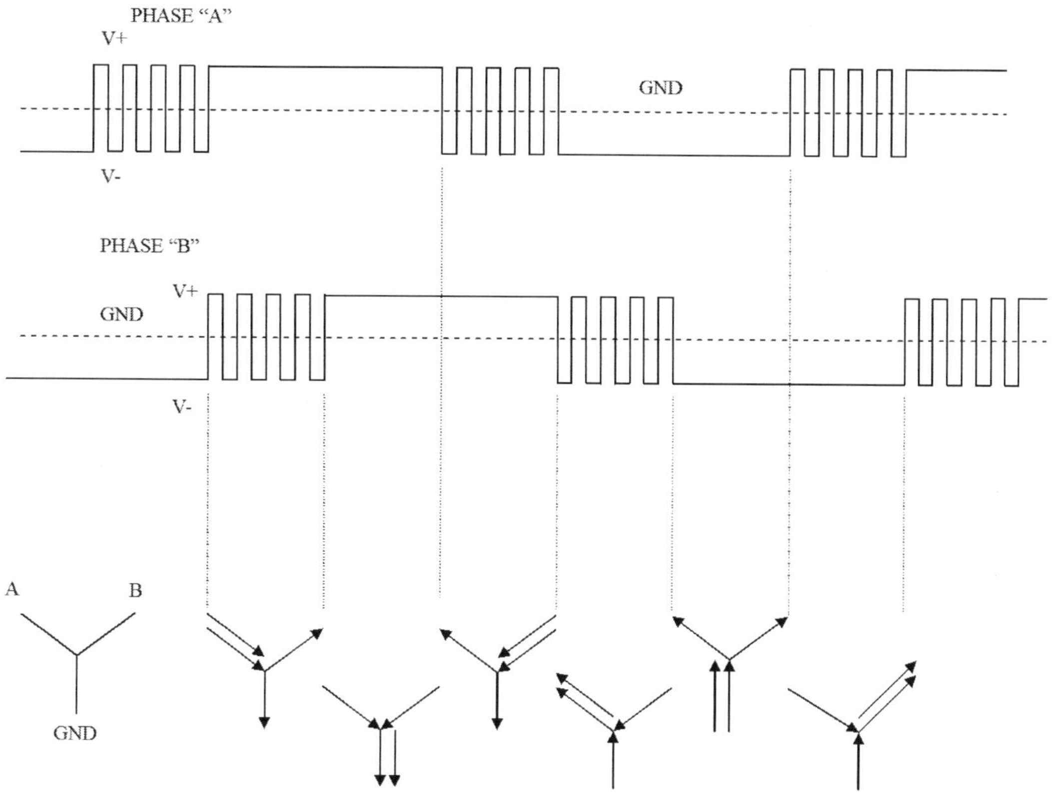

Differences between three phases and two-phases-plus-neutral: The major difference in the drive systems for a three-phase motor drive and a two-phase-plus-neutral motor drive is that both a positive and a negative voltage source is required for the two-phase drive. The three- phase drive requires only a single voltage supply, however producing both a positive and negative supply is quite simple when using a poly-phase or even a single phase A.C. power source.

The three-phase A.C. motor drives require six active components in order to produce a six-step waveform or a Pulse Width Modulated waveform at their output. The two-phase and Neutral can be created with four active components. For the two-phase-plus-neutral motor drive the voltage must be higher in order to obtain the same RMS voltage across all three motor phases.

Six Switch Three Phase Motor Drive Block Diagram (SW = Active Power Switching Component)

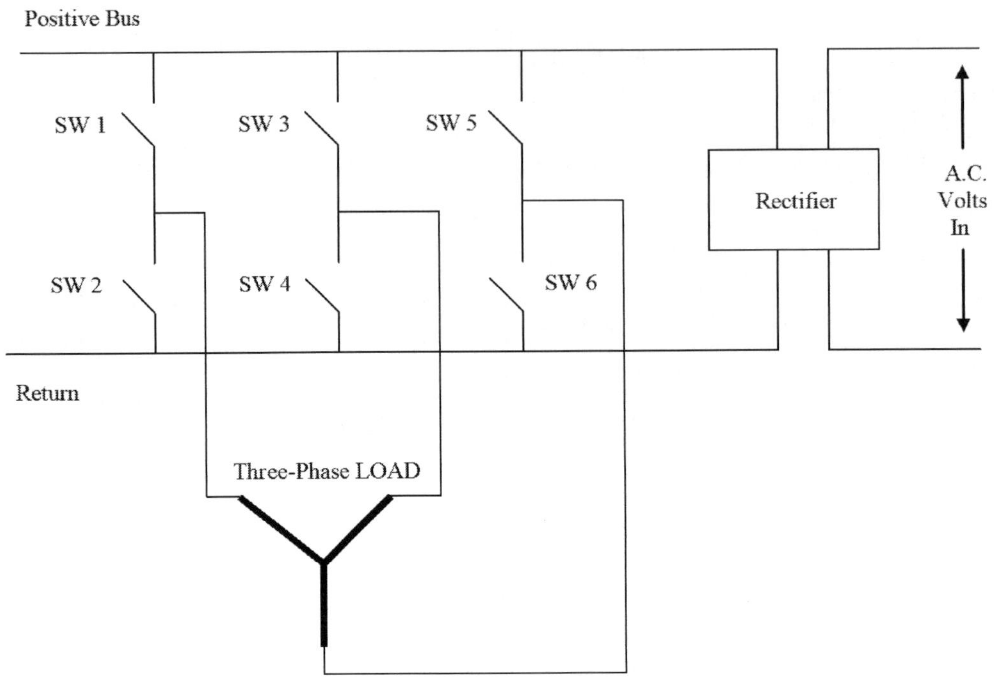

Four Switch Three Phase Motor Drive Block Diagram (SW = Active Power Switching Component)

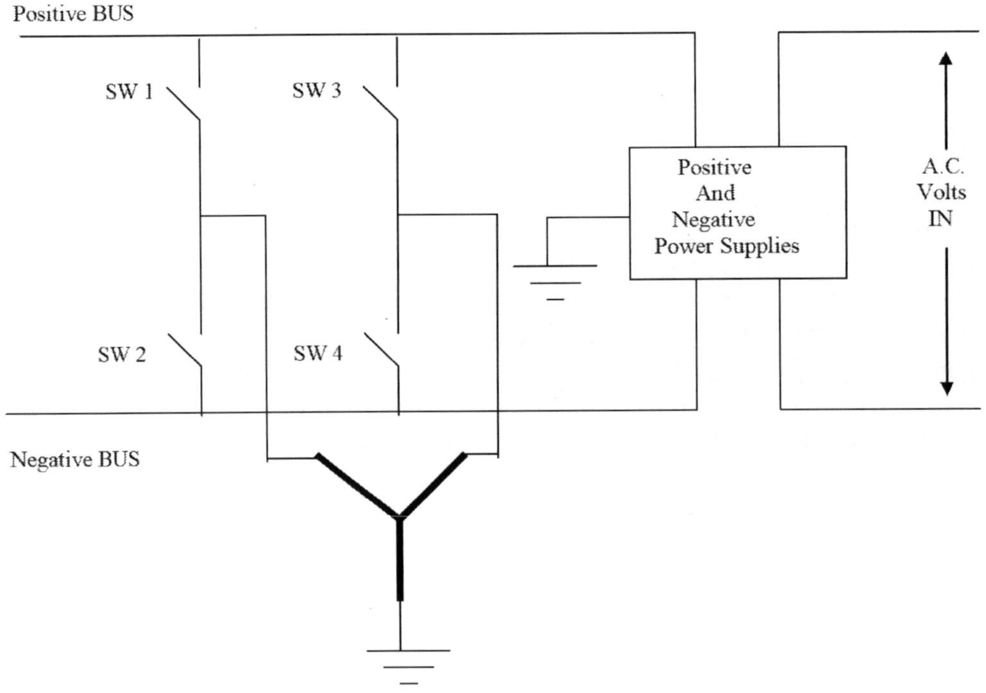

2-Phase Conclusion :

Using Neutral as one of the phases reduces the number of active components by one-third. These components are part of the more expensive components of any motor drive. A serious cost saving can be realized. A reduction in the number of power components also increases reliability.

Run with it.

Picking Your Jobs

I really have to say that since I stopped working for the corporate world and have been doing design work on my own I have never been happier. I can pick the jobs. I had thought that I could put as much time and effort into the job as I wanted. WRONG! Once you've accepted the job it kind-of owns you until it's finished or until it dies, or you die. The people you are working for are the same as the ones you worked for in the corporate world. They don't understand the design. They don't understand if it takes more time. They don't understand if it costs more than anticipated. They just want it done. Hey, can you blame them?

The best advice I have is to be honest with them. If they don't understand the design or why it's taking extra time or costs more, so what! If the results come within their window of acceptability they'll understand and probably give you other jobs. If they don't, chalk it up to experience and onward to the next customer.

The jobs I have accepted have always led to new experiences, sometimes with a new device, sometimes with a new approach to an old problem and sometimes with a whole new set of problems. Who can beat that? After all, engineers are problem solvers. There are a few things you should watch out for, though. As much as possible, make sure you understand what the customer wants. Don't expect a specification sheet, they know what they want. It's your job to read their minds as best you can. Things will most probably change as you get closer to having results. Plan on two or three…, O.K. o.k., three or more rounds of changes before you're approaching the end of the project. Next is not to try to perfect the thing. If the customer accepts, and is happy with what you have done, don't go overboard making improvements. It's not what the customer wants. You can go out on a limb a little bit if the customer wants something you're not quite sure of, but remember honesty? Let the customer know that it will take extra effort.

Most projects will have several aspects to them. Don't get bogged down doing just one aspect and especially, don't get bogged down doing only the parts that you like to do. It's easy to fall into this trap when you're the boss.

A Potentiometer Input Loop-Powered 4-20mA Circuit

So what is "loop powered". This term is used for 4-20mA current control devices that include circuitry capable of using the power in the loop that they are controlling.

What?
O.K., O.K. This takes a little thought. If the circuitry that controls some sensor has to be powered by the same power that it is regulating, then the worst case power into this control circuitry is the limiting factor. Since we are controlling a 4-20mA current loop, we know that the minimum current out of the loop at any time will be 4mA. So the control circuitry can use up to 4mA.

'What about Voltage?' you say. Well, as long as the input voltage level is sufficient to operate the control circuitry and to drive 20mA through the 'loop', that's all we need. In most current loops, the available voltage drive is in the range of 24 to 32 Volts. Let's assume that 24Volts is applied. If the control circuitry drops a maximum of 8 Volts with 24V applied, the voltage remaining to drive our current is 16 Volts. At our worst case of 20mA, the resistance of the loop can be as high as 16Volts/0.02A, or 800 Ohms. That's a lot of wire length or corroded connections. The idea is to measure every bit of current used by the control circuitry so that we can subtract this amount from the output current because the current through the control circuitry is already making it to the output. I know! I know, this is still confusing. Just remember that the control circuitry can't use more than 4mA.

So what does this control circuitry look like and how does it operate? The control circuit in the figure below, entitled **'Loop Powered Circuit'**, uses a voltage regulator on its' input. This regulator is sitting on top of a 2.5V reference and a 100 Ohm resistor, R3. Remembering that the maximum current that is available to the control circuitry is 4mA, if all of this current were passing through R3 we would only get a 0.4V drop due to R3. This would make the maximum voltage dropped by the reference, by R3 and by the regulator be 6.2 Volts. Because the regulator will require a little bit of voltage overhead in order to regulate (let's say 1.2V), the lowest voltage input at which our control circuitry will operate properly is 7.4V. Nuff said.

Why do we need R3 at all? Because we wish to measure the current through the reference pin of the voltage regulator and R3 provides a voltage drop that can be related to this current. What other currents do we need to measure? Well, what about the current out of the regulator.

Loop Powered Circuit

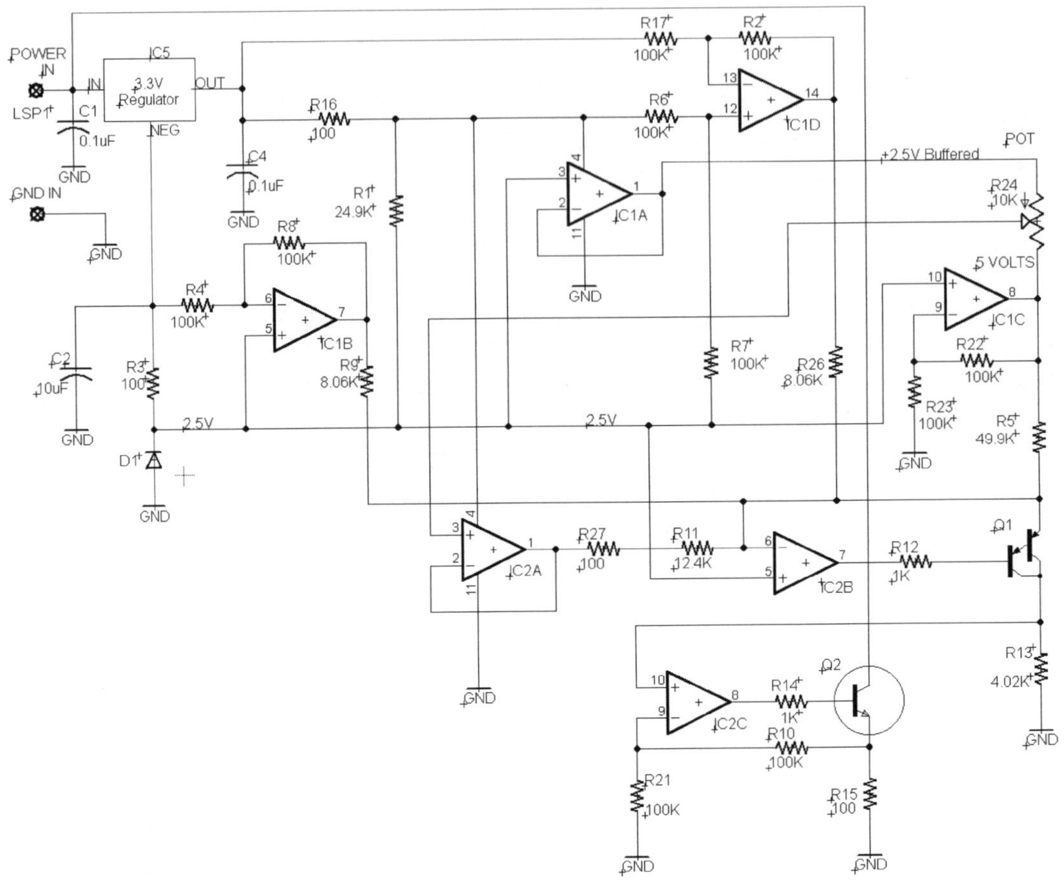

Do we need to measure the input to the regulator? Not really. We have measurements for the current through two of the regulator pins and the third pin must have the total of these two, just like a current node. With different input voltages the voltage regulator may drop different Watts, but this is accounted for in the two pins being measured. Where else should we measure? Well, since all current paths lead through the op-amps, and the op-amps are powered by the voltage regulator, all currents are accounted for.

What about R15? R15 is in series with the 'loop'. With 20mA through the loop minus the control circuit current that has been added to the loop without passing through R15, the worst cases voltage drop across R15 will be 100Ohms*0.02A or 2 Volts. This is well below the approximate 7 Volt input requirement for the control circuitry so the voltage drop across has no bearing on 'loop' current.

What does R15 do, anyway? Well, it is used to measure 'loop' current in terms of a voltage drop and to compare this to another voltage that represents 'desired loop current' minus the current already used by the control circuitry. IC2-C is used to make this comparison and outputs base current to Q2 that compensates for control circuit current. This base current is stolen from the 'loop' current and is included in the control circuitry current because it passes through an op-amp output. Still, this lost base durrent is a potential source of error so it is best to use a high-gain transistor, such as a Darlington, for Q2.

So how do we arrive at the voltage that represents 'desired loop current' minus the control circuitry current? We start with the 'desired loop current' setting. This, it turns out is the voltage on the wiper of potentiometer R24. This wiper voltage is buffered in IC2-A the output of which passes through a 12.5KOhm resistor to the '-' input of IC2-B. This connection is also part of a summing node. If the wiper of the pot is at the 2.5V end of the pot, there will be no contribution to the summing node. With the wiper set at any other position, a current will be added to the node.

There are three other currents passing through this summing node. These include a current that represents the current passing through the NEG leg of the 3.3V regulator, a current that represents the current passing through the 'OUT' leg of the 3.3V regulator and, last but not least a current that represents the 4mA constant minimum current. Here's the kicker. The feedback loop for IC2-B is also included in this node. IC2-B is configured as an inverting amplifier with a reference input connected to 2.5V. The feedback loop to the '-' input of IC2-B includes base drive resistor R12 and the base-emitter junctions of PNP Darlington transistor Q1. The voltage at the '-' input of IC2-B will be the same as the voltage on its '+' input, or 2.5Volts. This also happens to be the voltage of the node. What? How? Well, it looks like the only way this can happen is if Q1 is either not conducting at all or is taking away current from the node. That's exactly what is happening.

'Which of these is which?' you say. Well, the NEG current is measured as a voltage across R3. This voltage is inverted in IC1-B which is referenced to the 2.5V reference voltage. IC1-B's output is used to produce a current that passes through R9 to the summing node. Because the output of IC1-B is negative with-respect-to the 2.5V reference voltage, the current through R9 will be subtracted from the node. The same situation exists for the OUT current. This is measured first as a voltage across R16, then as a current through R26 which is connected between IC1-D and the node. IC1-D is configured as a difference amplifier that outputs the difference between the voltage across R16 as referenced to 2.5V. The

output of IC1-D is also inverted, so the output voltage will be negative with-respect-to 2.5V and the current through R26 will be subtracted from the node.
The third input to the node is the current through R5. IC1-C is configured as a non-inverting voltage doubler that doubles the 2.5V reference voltage, so the voltage at the output of IC1-C and therefore the voltage at the non-node end of R5 is a constant 5V.

Let's see how it works. Assume that there is no current through the IN and OUT terminals of the regulator. I know, I know! It's not a real-world situation. But for this example it is. With no current passing through the IN terminal of the regulator, there will also be no current through R3. With no current through R3 there is no voltage drop across R3. Since IC1-B is referenced to 2.5Volts The output of IC1-B will be 2.5Volts. With no current at the OUT terminal of the regulator there will be no current through R16 and thus no voltage drop across R16. Since IC1-D is referenced to 2.5V, the output of IC1-D will be 2.5Volts.

Let's also assume that the pot wiper is positioned on the 2.5V end of the pot. This voltage out of buffer IC2-A will be 2.5Volts. Consider that the node is satisfied. With the node voltage, and thus the '-' input of IC2B at 2.5Volts there will be no current through the R27-R11 series resistance.

Now let's look at R5. Remember that the 'R12 - Q1' feedback loop to the '-' input of IC2-B is trying to keep the node at 2.5Volts. If it is doing its job, there will be 2.5Volts at the node end of R5. The other end of R5 will have 5Volts on it. So the current through R5 is approximately 2.5V/50kOhms or 50 microAmps. This is the only current added or taken away from the node. Since this current is being added to the node, Q1 will have to conduct 50 microAmps to GND in order to keep 2.5Volts on the node. No problem. What happens to this 50 microAmps? It passes through R13 and produces a voltage drop of approximately 50uA*4kOhms or 0.2Volts. Then what! Well, this voltage is compared to half of the voltage that appears across R15, a 100 Ohm that is situated directly in the 20mA current loop. IC2-C controls the current through Q2 and forces the voltage drop across the 100 Ohm resistor to be twice this 0.2Volts or 0.4Volts. The resulting current through resistor R15, and thus through the current loop is 0.4/100 Ohms or 4mA. With only the current through R5 contributing to the current node, the loop current will be 4mA, its lowest value.

What about a situation where there is current through the regulator? Let's make an assumption that the current passing through the NEG pin of the voltage regulator is 1 milliAmp. Looking at the circuit, this will cause a 1mA*100 Ohm, or 0.1Volt drop across R3. What does this do to the OUT voltage of the regulator? The regulator will be re-referenced to 0.1V plus 2.5V or 2.6V above

ground. The regulator output voltage will be 5.9 Volts. Cool! But who cares? This output is rather meaningless to circuit current measurements. Let me explain. The current at the OUT pin will be whatever is required to run the op-amps and associated circuitry. If R16 measures this current, no matter what it is, then we don't need to worry about the regulator output voltage. And that's just what happens. R16 measures the currents passing through both op-amps.
A derivative of this current appears as a voltage at the output of IC1-D and causes a current through R26 which is subtracted from the node.

Now let's assume that the current through R16 is 2.2 milliAmps. The voltage drop across R16 will be 100Ohms*2.2mA or 0.22Volts. This voltage is inverted in IC1B and becomes 2.5V-0.22V or 2.28Volts. With the node end of R26 at 2.5Volts, current will be subtracted from the node. The amount of current subtracted is 0.22V/8.06kOhms, or approximately 27.3 microAmps. Remember this number.

What about the 1 milliAmp current through R3? The 0.1 Volt drop it produces is inverted in IC1-B and shows up on one end of R9 as 2.4Volts. The voltage on the node end of R9 is 2.5Volts, so the 0.1 Volts appears across R9 and the current through R9 will be subtracted from the node. This current is 0.1V/8.06kOhms or approximately 12.4 microAmps. How about the current to the node through R5? That remains the same. We still have 2.5V across R5 so a current of 50 microAmps flows into the node. The only current left is the current generated by the POT wiper position. For the time being, let's leave this wiper at the 2.5V end of the potentiometer.

So what's the total node current? Here we go. We have 50 microAmps into the node through R5. We have 27.3 microAmps leaving the node through R16 (I told you to remember this number) and we have 12.4 microAmps leaving the node through R3. This gives us a total of 10.3 microAmps entering the node. Remember Q1? This 10.3 microAmp current has to pass to GND, through Q1 (minus some very small base drive current). On its way it causes a voltage drop of approximately 0.0412Volts across R13. IC2 compares this voltage to twice the voltage across R15 and controls the base of Q2 to get this value. So the voltage across R15 will be 0.0824Volts. This produces a current of 0.0824V/100 Ohms or 802.4 microAmps.

That's it? That's all the current we get into the loop? How about adding up all the currents that are used by the control circuit. We had 2.2 milliAmps through R16 from OUT of the voltage regulator. We also had a 0.1V drop across R3, which computes to 0.1V/100 Ohms or 1 milliAmp. Let's total it up. 802uA + 1mA + 2.2mA = 4.02mA. Close enough!

So with the pot wiper set to the 2.5V position we wind up with the lowest loop current, or 4mA. What about moving the pot wiper to the other end so that there is 5Volts on the wiper. Now there will be a large contribution to the node from the current through R27 and R11. How much? Well, the voltage out of IC2-A will put 5 Volts on one end of the R27-R11 series resistance. The other end will be at node voltage, or 2.5 Volts. Therefore, the current through the R27-R11 series resistance will be 2.5V/12.5kOhms or 200 microAmps. This current will be added to the node.

If we use the same current values as in the last example the resulting node current will be +200uA from the POT setting, +50uA through R5, -27.3uA through R26 and -12.4uA through R9, or 210.3uA. Once again, this current flows through Q1 and produces a voltage drop of approximately 210.3uA*4kOhms or 0.841Volts. After IC2-C drives Q2 so that twice this voltage appears across R15, the resulting current through the loop is (2*0.841V)/100Ohms, or 16.8 milliAmps of loop current.

This doesn't look like anything I can relate to. But wait, we have forgotten that the circuit operating current has not been included in this calculation. Remember that the current through R3 is 1mA and the current through R16 is 2.2mA. My goodness! Add these to the 16.8mA and we have our 20mA, the maximum loop current. Now that looks familiar!

Setting the POT anywhere between its limits will produce a loop current that is somewhere between 4mA and 20mA. Here are some things to keep in mind. If we consider the current through R3 to be zero, then the voltage out of the regulator will be 2.5V+3.3V or 5.8Volts. The op-amps must run on 5.8Volts and be able to produce output voltages of 5V, like the output of IC1-C. Low current rail-to-rail input and rail-to-rail output devices are best. Also note that any change in node current produces a change in most of the currents and voltages. This circuit is a control loop in itself.

Fear nothing!

Dry Spells

There comes a time in every working person's life that the economy goes a little bit sour. O.K., sometimes a lot sour. These times can usually be foreseen to some degree and during that time it's a good idea to start looking at the job market to see what's available and what you're worth. Don't ever feel bad about looking on-line or in the papers to see what's in demand and what the going rate for your skills may be. I hate to tell you this, but you're a commodity. But, if you're taken by surprise, the best option is to just hang on. Don't jump ship with a rumor of the first ill wind. You selected (or tolerated) this job because at some point you liked what you were doing (or at least thought you did). Always remember, any job is better than no job. Another thing to keep in mind is that moving to a new job means that you're the 'new guy' and the new guy will usually get the axe before established personnel.

To continue, another hedge against the inevitable is to do a little moonlighting. O.K., it takes time away from the family and puts more pressure on you but you don't have to do a lot of it, just enough. In this way you're making new connections, and connecting is the name of the game. The old rules about being dependable, reliable and reasonable apply to moonlight jobs, just as they do with your regular job so don't overextend yourself. You'll wind up disappointing everyone. If you happen to lose your job, remember to read 'Caught in a Layoff' and 'Finding a Job' in this book.

Best of luck.

Safely Powering up 'H-Bridge' Circuits

One of the most annoying occurrences is to turn on power to your H-Bridge power drive circuit and to short the power source through one or both legs of the bridge. Depending upon the overcurrent protection of the power supply is not a good idea as this power supply may also run the lower power sections of the drive circuit such as the drive logic. This could create a very confusing trouble-shooting problem. Anyway, this circuit doesn't depend upon I.C.s in the drive circuitry that may pass through some quasi state as power is applied or removed, causing the output FETs to be on or off unpredictably. The components are limited to transistors, diodes capacitors and resistors. The purpose of the circuit is to keep the power components of the H-Bridge from turning on until all of the drive and control circuits have had a chance to stabilize and the power supply is either close to full potential during power-up, or close to zero during power-down. The diodes are low voltage drop diodes such as Schottky diodes.

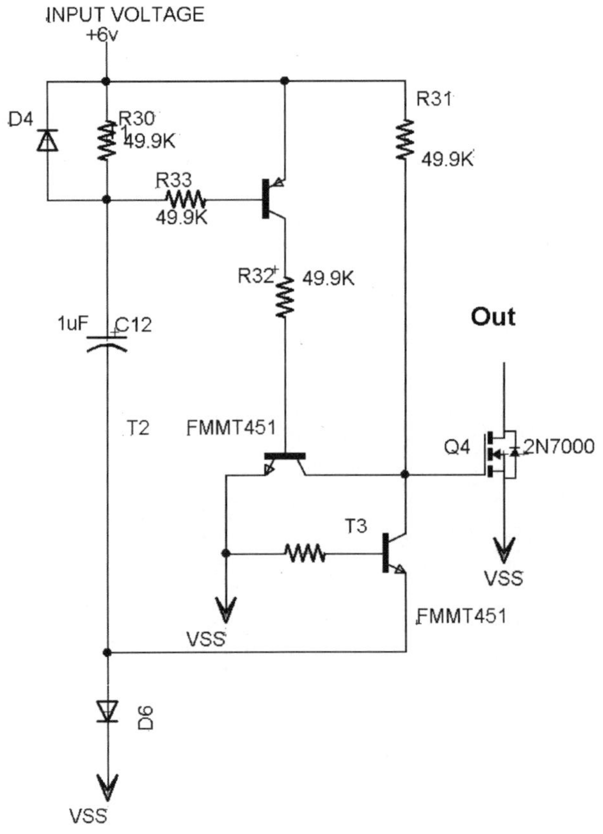

The circuit should operate from the same source as the Drive Logic or, if there is a voltage regulator between the incoming supply and the drive logic, the circuit can be operated from the incoming supply.

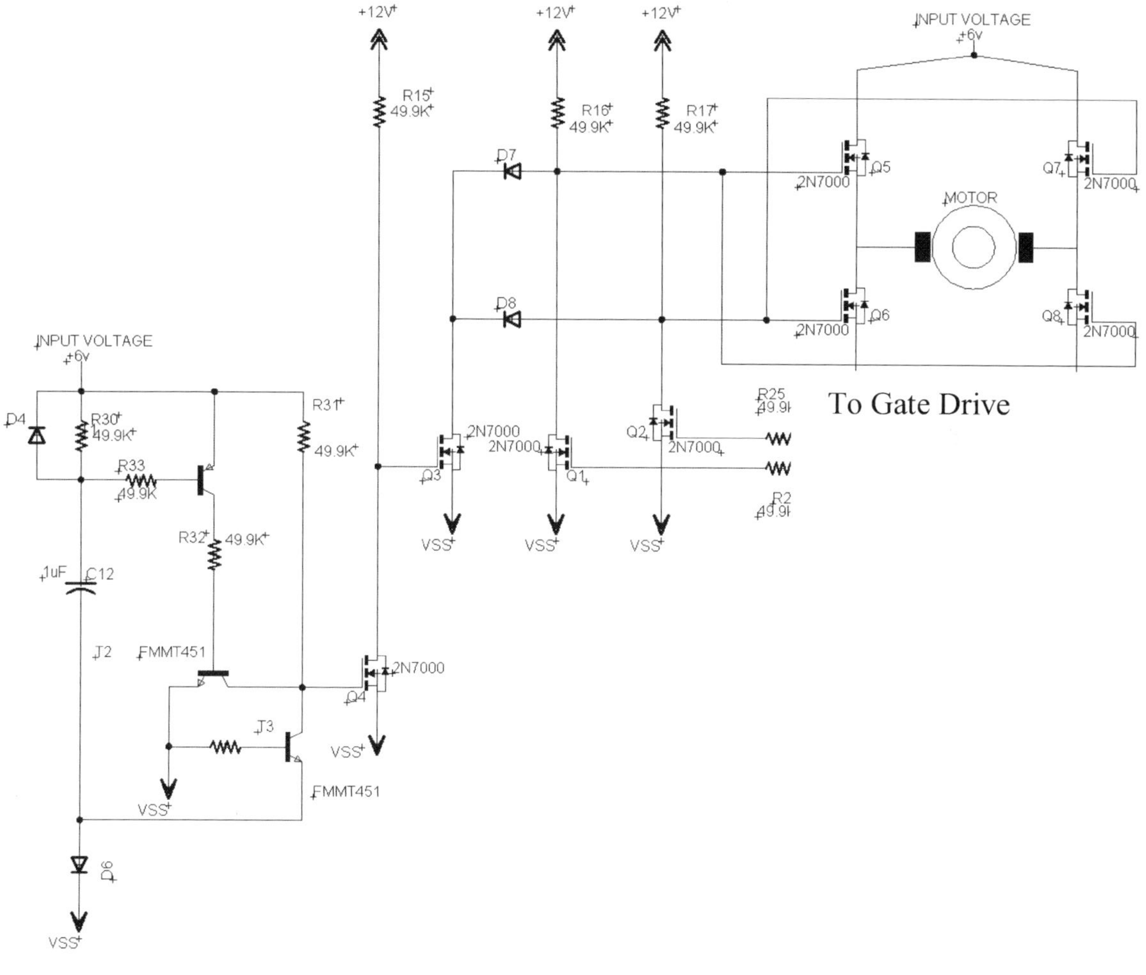

The fast action of this circuit is due to C12. Whenever C12 is charging or discharging, gate drive to the output devices will be cut off. When C12 is either charged or discharged there will be no effect on the gate drive from this circuitry and the Drive Logic will control the drive to the gate. The gate drive power source is not shown, but produces the '+12V' in the circuit above. It is used to power the H-Bridge drive components. That is why it is at a higher voltage level than the 'Input Voltage'. This allows the gate turn-on voltage to always be high enough to turn on transistors Q5 and Q7 of the H-Bridge.

What happens when power is applied? Let's assume that C12 is fully discharged at this time. As power supply voltage increases (nothing is instantaneous), C12 starts to charge through R30 and D6. D6 is forward biased and only provides a small series voltage drop. As soon as the Logic power supply exceeds approximately 1V, Q1 will begin to conduct. Current through Q1 flows through R3 and the base of Q2, causing Q2 to conduct. With Q2 conducting, there is current in the R6, D2, Q2 path and the R5, D3, Q2 path so that the voltage at the bottoms of R5 and R6 is approximately equal to 0.5V. In the circuit above, this will turn off FET Power Devices T4 and T5 as well as T3 and T6. C1 will continue to charge through both resistor R1 and through the path consisting of R2 and the emitter-base junction of Q1. When C1 is sufficiently charged, it causes Q1's base-to-emitter voltage to be less than 0.7V and Q1 will stop conducting. This also stops Q2 from conducting. The current through Q2 will decrease, allowing the voltage at the anodes of D2 and D3 to increase. This will allow the turn-on condition of T3,4,5 and 6 to be controlled, through T1 and T2, by the outputs of the Drive Logic.

How about the power-down condition? Assume the C1 is charged so that the voltage at the top of C1 is nearly at V+ potential. As V+ decreases (with respect to GND), the voltage at the bottom of C1 decreases. With V+ continuing to decrease, this voltage actually becomes negative with respect to GND. D4 becomes reverse biased and blocks the discharge of C1 to GND. The only other current path for C1 is through the emitter-base junction of Q3. With the base of Q3 tied to ground, the emitter voltage of Q3 cannot go below 0.7V so, as the voltage at the bottom of C1 tries to go more negative, the current out of C1 must increase in order to keep 0.7V at the bottom of C1. This increased current passes through Q3 and through R4. The voltage drop across R4 increases and the voltage at the bottom of R4 decreases to near GND potential. The cathodes of D2 and D3 are then at near GND potential and their anodes are about 0.3V above GND. The result is to prevent T3,4,5, and 6 from turning on. When C1 is fully discharged, no further current flows through Q3 and the cathodes of D2 and D3 rise to the voltage level of the Logic Power Source. This allows the Drive Logic circuitry to once again control the output FETs.

Are there any potential problems? Always! The circuit doesn't operate until the supply has reached an output of 1Volt or greater, or until the supply has decreased at least 1Volt below max. If the Logic Power Source is of the low voltage variety (3V or less), there isn't much of an operating range.

But it's better'n nothin'. Best of luck.

Pointing out the Obvious - Non-linearity of Parallel Resistances

This may seem like a silly time to present such a simple circuit, but it will serve as the introduction to another, slightly more complicated circuit. The whole deal is behavior of resistors in parallel, or just resistors in general. Two resistors in parallel follow the formula:

$$\frac{R1 * R2}{R1 + R2}$$

If one of these is a variable resistor and the other a fixed resistor, the resulting resistance across the two resistors follows a parabolic curve. Use two resistors, one fixed 10K Ohm and one variable 10K Ohm according to the schematic below:

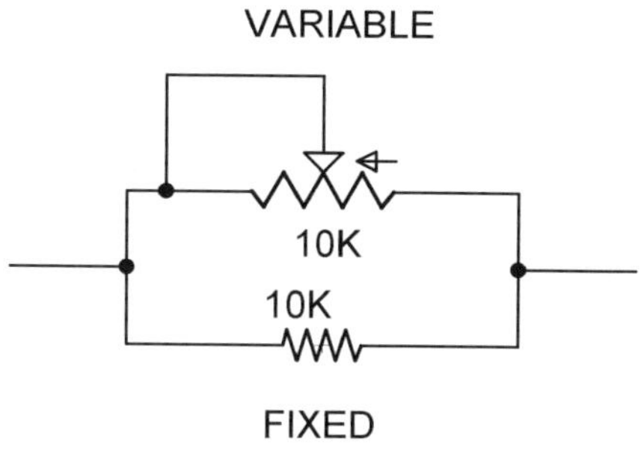

When the 10K variable is set at zero ohms, the resistance across the parallel combination is zero. When the 10K variable is set at maximum (10K Ohms) the resistance across the parallel combination is 5KOhms (100 / 20). How about when the variable resistor is set at 7.5K? Then we get slightly more than 4.28K Ohms. When it's set at 2.5K we get 2K Ohms. The plot of these points is::

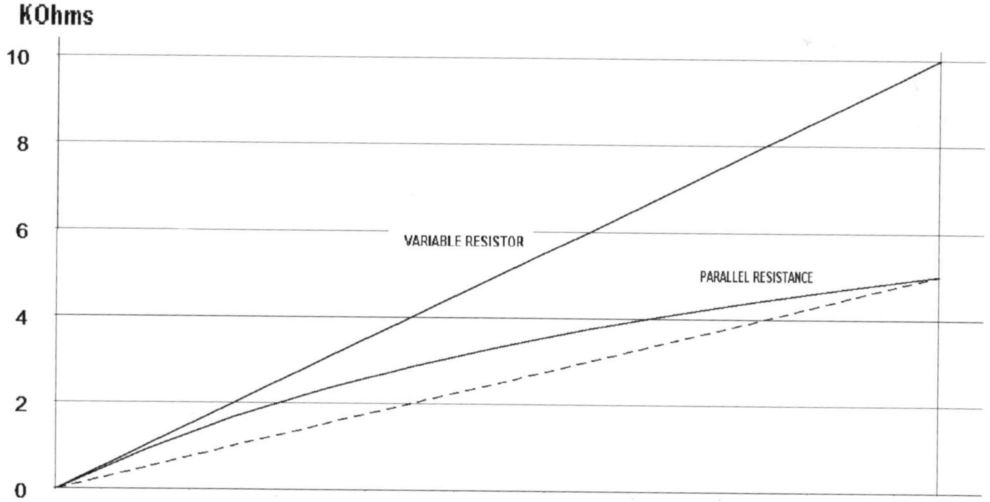

Because of the multiplication in the numerator of the parallel resistance formula, we wind up with a curve (a non-linear function).

Making Products Manufacturable

This is something that you have to keep in mind during the whole design process, especially for high-volume products. Aside from the normal operating specifications you should also be aware of how things are manufactured. Adding a connector or two may make things a lot easier than having to solder wires to your board. Stick them on the schematic so that it will be followed throughout the whole process. When your incoming power is in doubt, don't forget to use a regulator and don't forget the power supply reverse protection diode. If there is a plan for testing the circuitry, add test points or a test connector or two. If there is no plan for testing, make one! Then add test points, etc. as required. Here are a few general rules;

- Once you have a design, try to reduce circuit complexity.
- Try to reduce unit cost by keeping component values the same so that they can be purchased in higher quantities.
- Don't over-design but stay within the specs (customer specifications).
- Get manufacturing involved before you put pen to paper. Make sure that the design spec has been approved by everyone involved, especially the customer(s). Get it in writing!
- If mechanical and/or packaging engineering is involved, get the mounting locations, board shape and hole locations, weight, etc. O.K.ed 'in writing'.
- If you're the one writing the specs, check the working environment of the device:

i) Moisture tolerance
ii) Radio Frequency immunities
iii) Lightning susceptibility
iv) Power surges
v) Type of batteries, battery draw and battery life
vi) Display illumination
vii) Input and output protection
viii) Things you and I haven't thought of

Simulating a Variable Resistor with Active Components

Let's suppose that we want to simulate a resistor with active components. When would we want to do such a thing? Well, there are a lot of applications out there that still use potentiometers for control purposes. Of course, the main disadvantages of 'pots' is that they get 'dirty' or 'scratchy' (oxidized). The result is that the wiper loses contact with the resistive material. The contact can become so bad that it can 'open', leaving an infinite resistance value with unpredictable results.

Now comes the question 'exactly what does a resistor do?' Well, a fixed resistor serves to limit current in a circuit based upon its resistance value and the voltage that appears across it. It does this according to the formula I = V/R. Notice the division operation. We're going to need an electronic device that can divide. How about a multiplier.

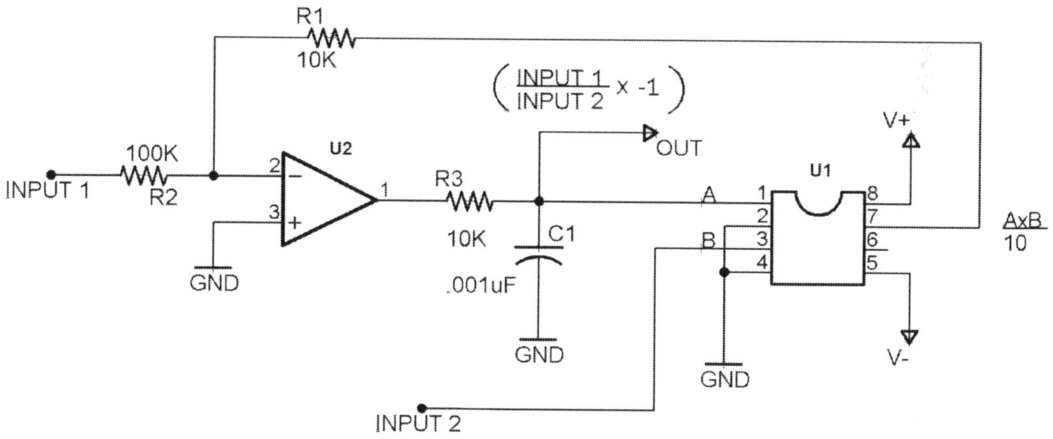

The multiplier in the schematic above is an Analog Devices 633 device. Almost all reasonably priced multipliers need a plus and a minus supply. All have outputs that are A*B/10. This allows the multiplication of 10V on INPUT #1 and 10V on INPUT #2, resulting in 10V out. So what is the configuration that turns this into a divider? Here's the deal. The question that needs asking is 'what input value to the multiplier has to be multiplied by a known input in order to get a known output'. As a 'known' input lets assume 2Volts on Input 2, above. Then let's assume 10 Volts is placed on 'Input 1'. We know that 10V divided by 2V is 5. The minus input of IC2A must be zero in order to satisfy the amplifier. With 10 Volts across R2, the current through R2 is 0.1mA (10V/100,000Ohms). This same current must pass through R1. The resulting voltage drop across R1 is 0.1mA x 10KOhms or 1 Volt. Because the current is flowing from Pin 2 of IC2A through R1 and the Pin 2 end of R1 is at Ground potential, the voltage at the multiplier end of R1 must be negative. Therefore, -1V appears at the output of multiplier pin 7.

Now we have the multiplier 'known output' value. The last thing we need is the multiplier input voltage value ('A') on pin 1 of the multiplier. 2 volts on input 'B' multiplied by 'x' volts on input 'A' and then divided by 10 gives us -1 volt out. So input 'A' must be negative. How negative? Well, it's;

$$\frac{(2 * x)}{10} = -1, \quad 2 * x = -10, \quad x = -10/2 \quad \text{or} \quad -5\text{Volts}$$

Here's a circuit representation that puts it in mathematical terms (dividend, divisor and Quotient):

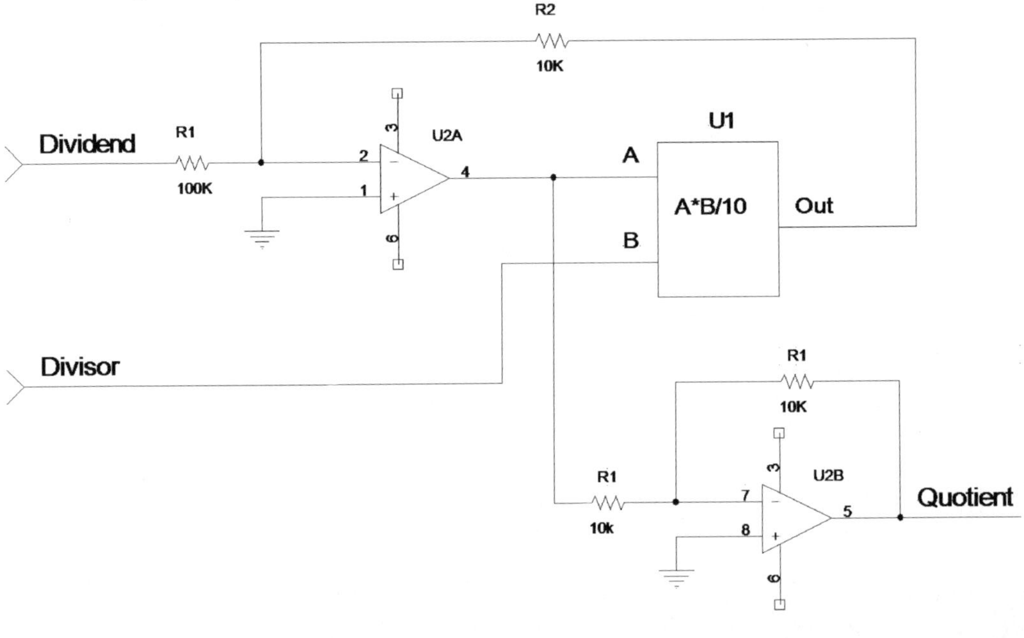

That's the divider portion of the circuit. Remember, we're trying to simulate a resistor. That means that one input to the divider will be the voltage across the resistor (the numerator of the 'V/I' term) and the other input will be a voltage representing the current through the resistor (the denominator of the 'V/I' term). So let's figure out what's needed for the circuit. We're going to need a device that limits current and is controlled by voltage in order to simulate a resistor. How about trying a transistor?

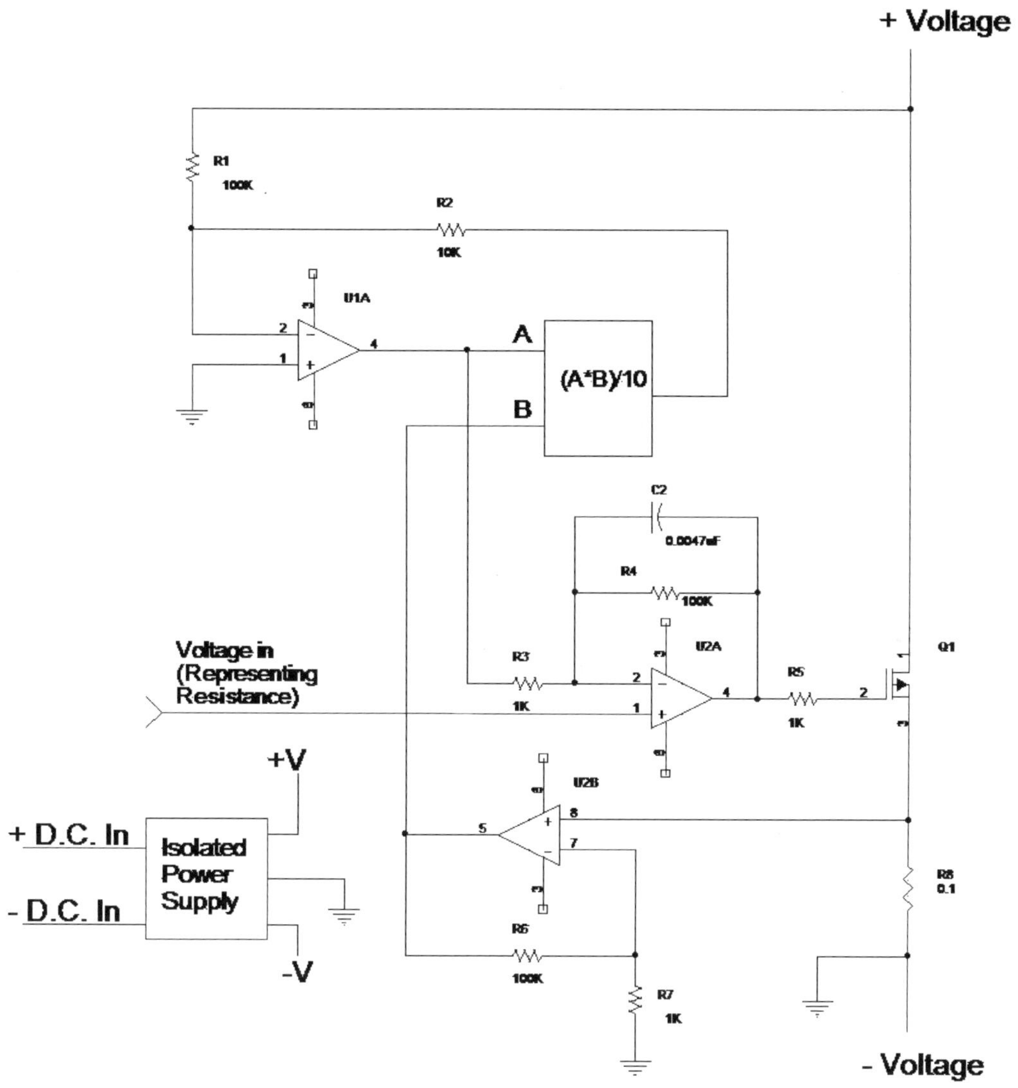

Hey! What's this? Suddenly we have some kind of nightmare circuit with feedback loops all over the place. Well, Q1 is there to act as our resistor and U1A, and the multiplier are there as we described before. What about U2A? It looks like a very high gain inverting amplifier. When it grows up, maybe it'll be a comparator. In this circuit arrangement, U2A is used to compare the 'A' input of the multiplier to some desired voltage level. The output of U2A turns on Q1 sufficient to have the product of the voltage across Q1 and the current through Q1 produce a voltage at the 'A' input to the multiplier that matches the desired 'Voltage in'.

Note that the current sensing resistor, R8 is an actual resistance in series with the variable resistance of Q1, so this will be the minimum resistance of the circuit, although Q1 will also have some minimum resistance even when fully on.

So here's the deal. The 'B' input to the multiplier is a voltage that represents the current through the resistor (active resistor, Q1). Input 'A' of the multiplier is a voltage that represents the voltage across the resistor Q1. Let's assume that we supply a constant 10Volts across the resistor (Q1). This means that the only variables will be the 'Voltage in', which represents the desired resistance value, and the current through Q1, represented by a voltage drop across R8.

So let's work backwards (just like the divider circuit that uses a multiplier). What is the 'Voltage in' value that produces a resistance of 100 Ohms? Now that we have assumed a 100 Ohm resistor, we know that the current must be 10V/100 Ohms, or 0.1 Amps. 0.1 Amps (100mA) through R8 produces a voltage drop of 0.1Ohms * 0.1 Amps or 0.01 Volts. This voltage value is multiplied by 100 in the U2B circuit and the output of U2B (and thus the 'B' input to the multiplier) is 1 Volt. Remember that we have assumed 10 Volts across resistor Q1. In order to satisfy U1A, the output of the multiplier must be -1V. With 1 Volt in the 'B' input, the only value at the 'A' input of the multiplier that produces -1V out is -10Volt [(-10V * 1V)/10]. In order to satisfy the high-gain amplifier U2A, the inputs to U2A must be almost identical. Otherwise, the output of U2A will cause Q1 to be either full 'on', or full 'off'. With -10V on the '-' input of U2A the voltage on the '+' input will effectively be -10V.

Let's try this one more time only we'll define the 'Voltage in' value first. Let's assume it to be -2Volts. If we continue to assume that the voltage across resistor Q1 is 10 Volts, the output of the multiplier will have to be -1Volts. With the output of U1A at -2V, the 'B' multiplier input must be 5 Volts in order to get -1Volt out of the multiplier [(-2V * 5V)/10]. 5 Volts out of U2B means that there are 0.05 Volts across R8 (5/100), so the current through R8 is 0.05/0.1 or 0.5Amps (500mA). So what's the resistance value of the circuit? Why it's 20 Ohms (10V/0.5A). So given two points (100 Ohms and 20 Ohms with 10V and 2V 'Voltage in' values, respectively) the output resistance appears to linearly follow the 'Voltage in' voltage values. What about C2? Well, without an 's' value component, or something to shift phase or slow things down in the circuit, an oscillatory condition could result. Remember, nothing is instantaneous.

What about the 'Isolated Power Supply'? This is required because we don't know what the applied voltage will be (we selected 10V, but it could be anything) and we don't want the added current draw of the circuitry to affect this voltage. The power supplies are '+' and '-' and the neutral is connected to the bottom of resistor R8. Speaking of R8, what about the Wattage of R8? If we assume that Q1 can become a perfect short, then the required wattage for R8 when zero Ohms is selected will be E^2/R. or 100V/0.1Ohms. This turns out to be 1000 Watts..Whoops!!!! So we're going to have a lower limit for the resistance value of the circuit. We're going to need a 'Watts vs. Resistor Voltage' curve:

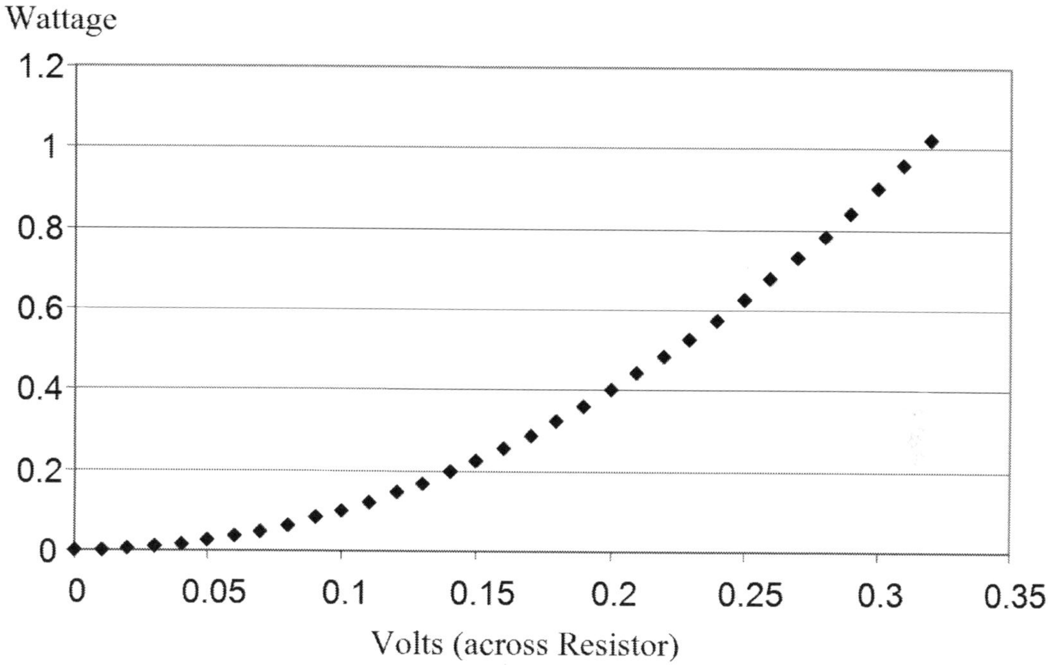

The above is a plot of Wattage being dropped by the resistor versus the Voltage across the resistor (with a constant resistance of 0.1 Ohms). The chart reveals that the Wattage dropped by the resistor is at 1 Watt when the voltage across the resistor is at approximately 0.32 Volts. The current at this voltage drop is 3.2 Amps. Therefore the resistance of Q1 must be no less than ' (10V – 0.32V)/3.2 Amps' , or approximately 3 Ohms. Not too bad! But Wait! This means that the resistor Q1 will have to drop 33 Watts (E^2/R or $10^2/3 = 33$W). That's a lot of heatsink. You may want to do another calculation with limited Q1 Wattage (the Voltage drop of the resistor is now quite insignificant compared to the drop of Q1. Just ignore it during Q1 Wattage calculations).

Another thing to notice is that Q1's gate is being driven by the output of U2A through a small resistance (1KOhm). Check the specs for Q1 and make sure that the negative voltage value of the gate is not exceeded.

One last thing. Notice that the above circuit only handles circuits with current that flows in one direction. To make it a 'real' resistor simulator, the circuit has to be duplicated for the opposite direction. Remember, the FET, Q1 has an internal diode that would be forward biased if the voltage applied across Q1 were reversed. This path can be blocked by means of a diode. Remember, you'll need another diode in the other direction if a duplicate but opposite circuit is used, and you'll also need another isolated power supply with its ground referenced to the other end of the resistor. Watch out for circuits that contain storage elements such as capacitors. Their charge or discharge paths may not be limited in the reverse direction when using the above circuit.

Watch Out for the Entrepreneur

Here's the thing. You like to design electronics. Someone, who you probably already know has an idea about starting a business or developing a product. They want you to design and build the prototype and then they'll market the product and you'll both share in the profits. Remember one thing. You're the one that first has to invest all of the time and maybe some, if not all of the money. They don't have to work a lick until you have put a lot of blood, sweat and tears into their venture. Even the simplest designs have a learning curve. In the beginning they'll probably stay involved in your design work by making added requirements of the product. These just take more of your time (and money) and that's more time before they even get productive. Nine out of ten times that time will be never.

Try to pick an entrepreneur who has been through this before and don't forget to ask a lot of questions before you lay pen to paper. I know, I know you'll probably be thinking of how to design the thing as soon as they approach you with their idea, but try to keep the emotion level down. It's hard because you probably love to design and to solve problems. Believe me, you'll be wasting your time. Make sure you outline the whole project with the amount of your involvement AND with the involvement of that entrepreneur guy. Once you've agreed upon the plan, only an act of congress should be able to change any part of it. Make sure you have the time available and always remember that after you have an estimated time-to-completion, that time should be tripled and the resulting time should then be doubled. After that, you'll probably still be in the hole for time.

Of course, if the entrepreneur is footing the bill on everything, forget what I just said. Most of the time that's not the case and when they have lost interest, you'll have put out a lot of hard work for no.

Take the high road. Ask the questions up front.
Good luck.

Non Inverting Op Amp Circuit with Gain and Offset

Sometimes you are left with one op amp and you need to do three functions, such as a non-inverted output signal, with gain and offset of some input signal. With limits, here is one way of satisfying these functions. What limits? Well, the input-to-output must not be inverted and the input-to-output range must have a gain of more than 1.

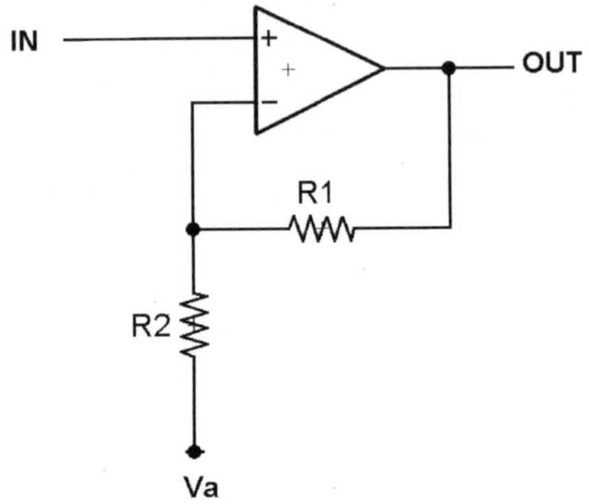

What about an example? Select some parameters that conform to the above limits. O.K.! O.K.! How about an input range from 0.5 volts to 2.0 volts and an output range from 3 volts to 9 volts. How are we going to relate these pairs of values (0.5V in yields 3 volts out and 2.0V in yields 9 volts out)?
Let's do a little math:

With 0.5V in:

$$I_1(\text{current through R1 and R2}) = \frac{3V\ (\text{op-amp output}) - 0.5V\ (V\ in)}{R1}$$

$$\text{while} \quad R2 = \frac{0.5V - Va}{I_1}$$

With 2.0V in:

$$I_2 = \frac{9V - 2V}{R1}, \quad \text{while} \quad R2 = \frac{2.0V - Va}{I_2}$$

At this point, let's arbitrarily give R1 a value of 1Ohm. We can do this because op-amp gain doesn't depend upon resistance values, just the ratio of resistances. Also, we are dealing with an ideal op-amp (sources or sinks infinite current). Ha! Plugging this value into the equation for I_1 we get:

$$\frac{3 - 0.5}{1}, \quad \text{or } 2.5\text{Amps.}$$

Plugging this value into the equation for I_2 we get: $\frac{9 - 2}{1}$, or 7Amps

Re-visiting our equations for R2 we can set the right-hand sides of each equation equal to each other and get:

$$\frac{0.5V - Va}{I_1} = \frac{2.0V - Va}{I_2}$$

Plugging in the values for I_1 and I_2 we arrive at:

$$\frac{0.5V - Va}{2.5} = \frac{2.0V - Va}{7}$$

Cross multiplying yields;

$$3.5 - 7Va = 5 - 2.5Va$$

Solving for Va; $= -0.33333V$

With this Va value, the circuit for the low input value now becomes:

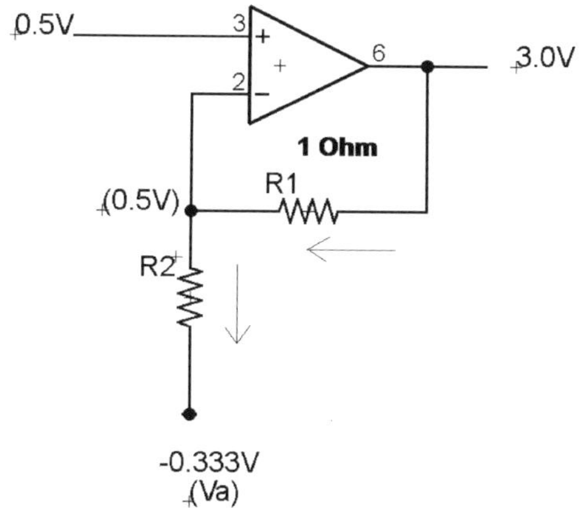

We can now determine the value of R2 having previously calculated the current through R1, which is the same current that passes through R2. The value of R2 is:

$$R2 = \frac{0.5V - (-0.333V)}{2.5A}, = 0.333 \text{ Ohms}$$

Make a quick check of the maximum input and output voltage ranges:

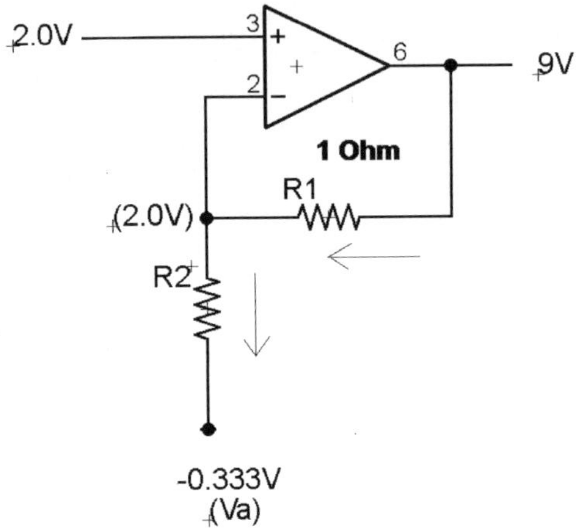

$$\frac{9V - 2V}{1 \text{ Ohm}} = 7 \text{ Amps} \quad \text{through R1 (and therefore, through R2)}$$

The value of R2 calculates to be: $\frac{2V - (-0.333V)}{7\ A} = 0.333$ Ohms

Which checks out.

Now it will become clear why we set R1 equal to 1 Ohm. Remember how we said that the gain of an op-amp is only dependent upon the ratio of resistances about the circuit? Let's determine the ratio between R1 and R2:

When R1 = 1, R2 = 0.333

$\frac{R1}{R2} = 3.0$, so the proportion is R1 = 3R2

Knowing this, we can make R1 equal to 10KOhms or 100KOhms. Then R2 becomes 30KOhms or 300KOhms. This keeps resistor values and circuit currents in a more reasonable range and allows us to use lower resistance values for a voltage divider to create 'Va'. Keep this Va current source/sink at a greater than 100:1 current ratio for 1% or less accuracy.

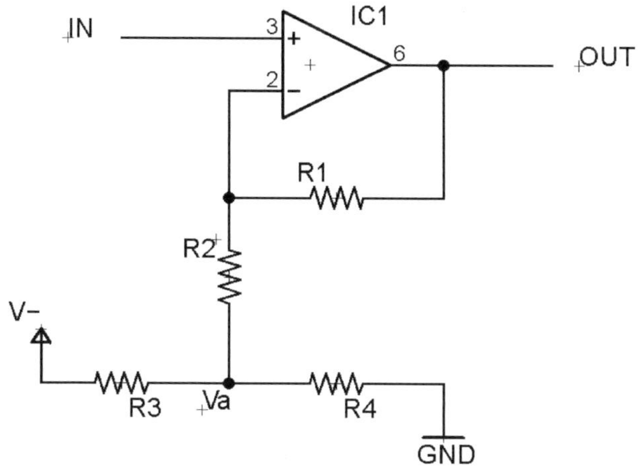

In the above example the current through R1 and R2 was always into the junction of R3 and R4. With other input-output ranges the op-amp will both source and sink current. This has to be taken into account when determining the values of R3 and R4. Let the value of R1 be 100KOhms and the value of R2 be 300KOhms. With Va equal to
-0.333 volts and the op-amp output at it's maximum of 9V, the current through R1 and R2 is:

$$\frac{9V - (-0.333V)}{300KOhms + 100KOhms} = 23.33uA$$

This current will be sunk at the junction of R3 and R4. In order to keep Va from shifting less than 1%, the current through R3 and R4 must be at least 100 times greater than the 23.33 uA x 100 or 2.333 mA. Using Ohm's Law:

$$R4 = \frac{E}{I}, = \frac{-0.333V}{.002333} = 143 \text{ Ohms}$$

$$R5 = \frac{E}{I}, = -\frac{12V - (-0.333V)}{0.002333} = 5{,}000 \text{ Ohms}$$

For less percent error, decrease the resistor values proportionately but remember we're using up 2.333mA in the above example and milliAmps may be at a premium.

So much for that!

Difference Amp for High Voltage Inputs

This circuit is interesting. I may not answer all questions about it but you'll see that it has some interesting component value calculations. The purpose of this circuit is to determine the difference between two voltages that far exceed the input voltage of the op-amp. The directions of current for i3 is correct because Va will be at a positive potential and OUT will be at a negative potential.

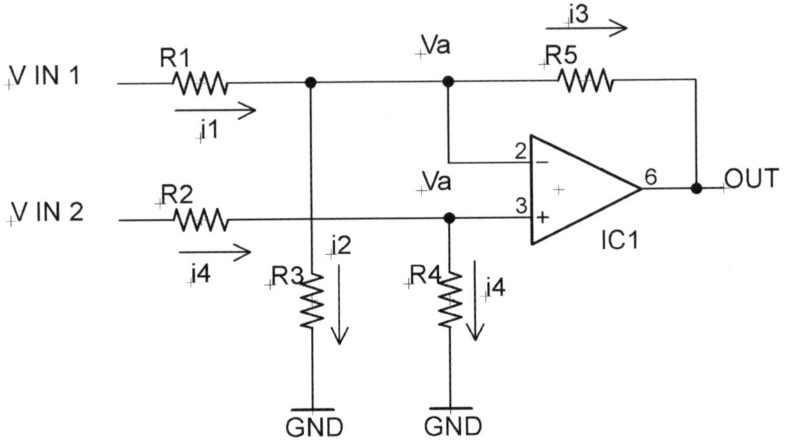

Let's pick a set of values for the circuit with a couple of assumptions. Assume that the values of R1, R2 and R5 are all the same and assume that the input voltage doesn't exceed 100 Volts. We're going to need fairly high resistance values to minimize any current from 100 Volt inputs to GND, so we'll select a value of 90kOhms. Now we'll select a value for R4. We don't want the voltage at the inputs of the op-amp to exceed the op-amp specs, so to be safe we'll keep the voltage value of Va at or below 10 Volts. A value of 10kOhms for R4 satisfies this condition;

$$Va = i4 * R4, = \frac{V\ IN\ 2}{R2 + R4} * R4 = \frac{99\ Volts}{100kOhms} * 10kOhms = 9.9\ Volts$$

What about the resistance value of R3? We're going to have to do a little math to determine this value. To keep things simple, let's assume that V IN 1 and V IN 2 have the same value. In that way we will be looking for a value of <u>zero volts</u> at 'OUT', the op-amp output. Also, let's put the value of R4 in terms of the value of R1 (or R2) by dividing R1 by R4;

$$\frac{100kOhms}{10kOhms} = \frac{R1}{R4}, \qquad \mathbf{R4 = \frac{R1}{10}}$$

We can represent i4 in terms of R1 as follows:

$$i4 = \frac{Va}{R4} = \frac{Va}{\frac{R1}{10}}$$

Now, what do we know about the circuit currents? Well, we know that i1 and i4 must be the same because the voltage drops and resistances of R1 and R2 are the same;

$$i1 = i4 = \frac{V\,IN - Va}{R} = \frac{Va}{\frac{R1}{10}}$$, where V IN1 = V IN 2 = V IN and R1 = R2 = R

Also, with OUT = zero volts:

$$i3 = \frac{Va - 0}{R5}, \quad i2 = \frac{Va}{R3} \quad \text{and} \quad i1 = i4 = \frac{Va}{\frac{R1}{10}}$$

We also know that $i1 - i2 - i3 = 0$

Substituting and solving for R3;

$$\frac{Va}{\frac{R1}{10}} - \frac{Va}{R3} - \frac{Va}{R1} = 0, \quad \frac{1}{\frac{R1}{10}} - \frac{1}{R3} - \frac{1}{R1} = 0, \quad \frac{1}{R3} = \frac{1}{\frac{R1}{10}} - \frac{1}{R1}$$

$$\frac{1}{R3} = \frac{10}{R1} - \frac{1}{R1}, \quad \frac{1}{R3} = \frac{9}{R1}, \quad \mathbf{R3 = \frac{R1}{9}}$$

With the given values:

$$R3 = \frac{100{,}000}{9} = 11{,}111 \text{ kOhms}$$

So let's see if it worked. Let's make the value of V IN 1 equal to 100 Volts and the value of V IN 2 equal to 99 Volts. Therefore, we expect an output voltage of -1 Volts. With R1, R2 and R5 equal to 100kOhms, R4 equal to 10K and R3 equal to 11,111K, the circuit looks like this:

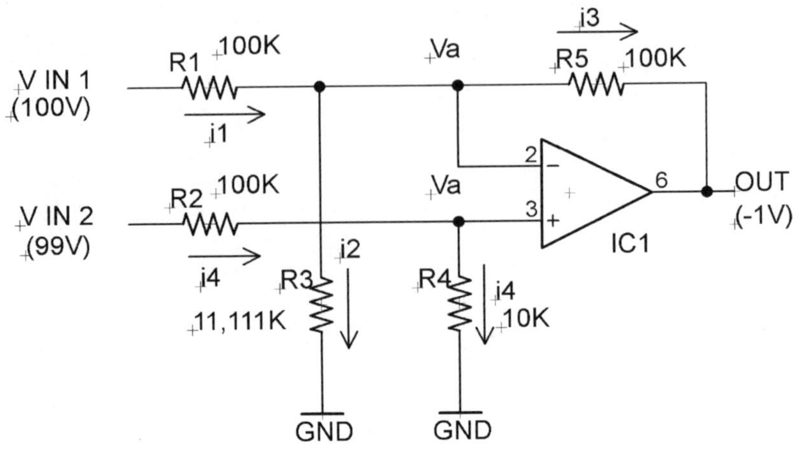

The value of Va will be; $\frac{\text{V IN 2}}{100K + 10K} * 10K$, $= \frac{99V}{100K + 10K} * 10K$

$= \textbf{9 Volts}$

Current i3 is;

$i1 - i2$, $= \frac{\text{V IN 1} - Va}{R1} - \frac{Va}{R3}$, $= \frac{100V - 9V}{100K} - \frac{9}{11{,}111 Ohms}$

= 99.999 micro-Amps, or 100uA

The voltage drop across R5 will be 100uA * 100KOhms or 10 Volts.
Since the voltage out of the op-amp is negative, the direction of current through R5 will be from left to right. So the voltage drop will be Va – 10V or;

9V – 10V = -1V, the value of the voltage at OUT

This checks out O.K.

Now for something very interesting.
Change the value of R4 to $\frac{R}{9}$ or 11,111 Ohms

Re-calculating the value of R5 as we did above, we first solve for Va;

$$Va = \frac{99V}{100K + 11{,}111 Ohms} * 11{,}111 \text{ Ohms} = 9.9V$$

Leave the input values (100V and 99V) and the output value (-1V) the same and re-calculate the current through R3

$$i3 = \frac{Va - OUT}{R5}, \quad i2 = \frac{Va}{R3} \quad \text{and} \quad i1 = i4 = \frac{Va}{\frac{R1}{9}}$$

We also know that $i1 - i2 - i3 = 0$

Substituting and solving for R3;

$$\frac{Va}{\frac{R1}{9}} - \frac{Va}{R3} - \frac{Va}{R1} = 0, \quad \frac{1}{\frac{R1}{9}} - \frac{1}{R3} - \frac{1}{R1} = 0, \quad \frac{1}{R3} = \frac{1}{\frac{R1}{9}} - \frac{1}{R1}$$

$$\frac{1}{R3} = \frac{9}{R1} - \frac{1}{R1}, \quad \frac{1}{R3} = \frac{8}{R1}, \quad \mathbf{R3 = \frac{R1}{8}}$$

With the given values: $R3 = \frac{100,000}{8} = 12,500\text{Ohms}$

So current i2 is now:

$$\frac{Va}{R3} = \frac{9.9V}{12,500 \text{ Ohms}} = 0.792mA$$

$$i3 = i1 - i2 = \frac{100V - 9.9V}{100K} - 0.792mA, \quad = 0.901mA - 0.792mA,$$
$$= 0.109mA$$

The voltage drop across R5 is 100,000Ohms * 0.109mA or 10.90V

Current still flows into the output of the op-amp so we subtract 10.9V from Va;

9.9V (Va) -10.9V = -1V OUT

This is the expected output.
But Hey! Notice the difference between the R4 and R3 with respect to R. In one case the ratio of R4 to R3 is; $\frac{R}{10}$ to $\frac{R}{9}$ (where R is equal to R1 or R2)

In the second case the R4 to R3 ratio is; $\frac{R}{9}$ to $\frac{R}{8}$ (where R is equal to R1 or R2)

This ratio holds (i.e. $\frac{R}{8}$ to $\frac{R}{7}$, $\frac{R}{7}$ to $\frac{R}{6}$, … down to $\frac{R}{1}$ to $\frac{R}{0}$) because

'R divided by 0' gives us infinity for the value of R3. This gives us a circuit that looks like a standard Difference circuit so look out for the voltage to the inputs of the op-amp.

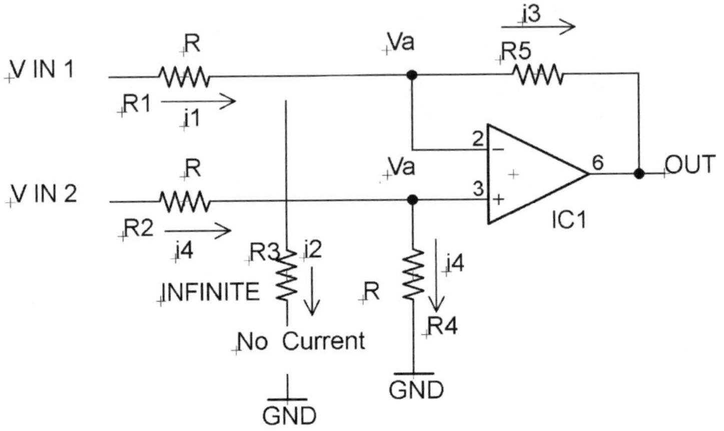

By the way, The proportionality to R doesn't have to be a whole number.

If you use $\frac{R}{8.37}$ to $\frac{R}{7.37}$ or anything like this, that works too!

One more thing. The op-amp doesn't have to be in the inverting configuration. You can also use a non-inverting configuration and calculate the R4:R3 proportions in the same way except that, in the calculations, the R3 value becomes the R4 value and vice-versa. The circuit below shows this swap.

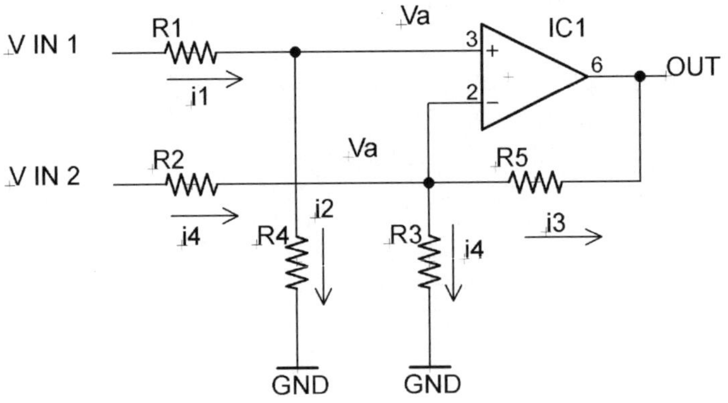

(Whew! I'm sure that dead horse we've been beating is very dead by now). Keep thinking.

An Analog Bi-Directional Switch

This may not be the most exciting circuit but it has come in handy on a few occasions. The device of choice for use as a switch, is the FET. Two parallel FETs, one conducting in each direction should be ideal.

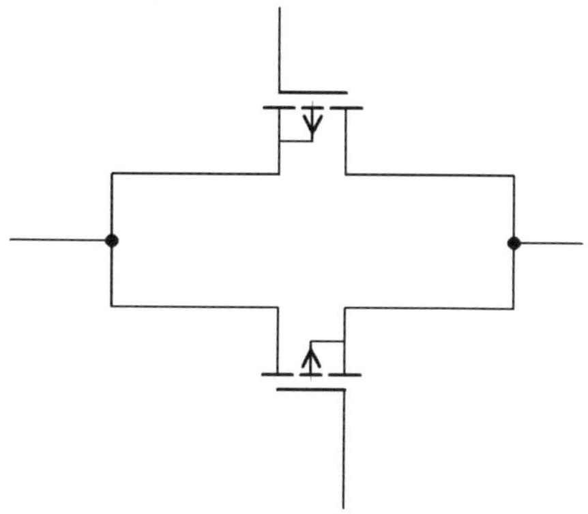

Here's the thing. When you go bi-directional with your signal, current has to travel through the circuit in both directions and a FET has a built-in diode in the reverse direction. So the circuit looks more like this;

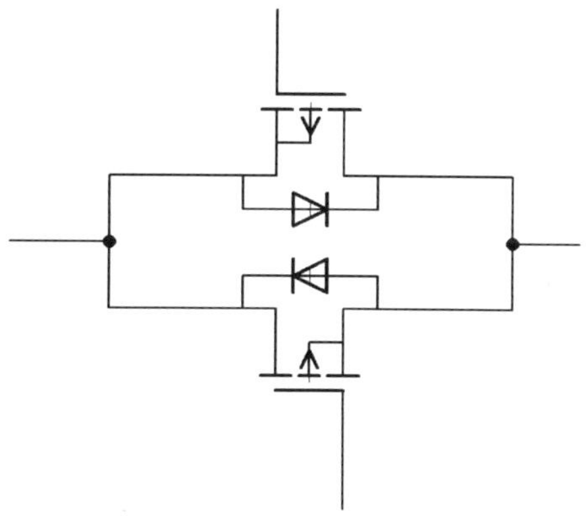

These diodes will conduct when they are forward biased. This occurs when they have approximately 0.7 volts across them. The signal voltage will be clamped at +/- 0.7 volts. How do we get around this? Well, the simplest way is to make the range of the input signal voltage so that it never exceeds +/-0.7 volts. A resistor divider serves this purpose nicely.

Assume a bi-polar power supply for the op-amps in the circuit below. Also assume an input signal (IN) that ranges from +5 to -5 volts. Also assume an On/Off Control Signal that can be -10 volts, +10 volts, or zero volts.

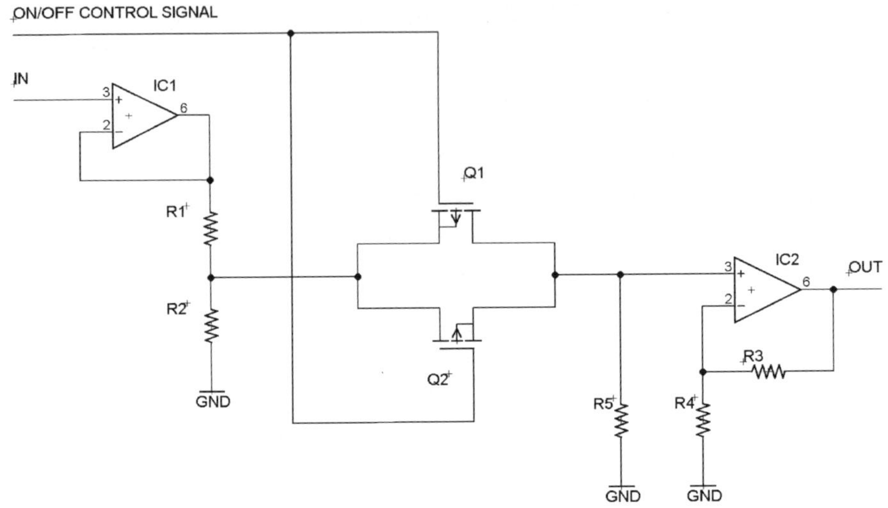

The voltage divider made up of R1 and R2 will reduce the input voltage so that it is in the 0.7V range. Just to be safe, make the range +/-0.5volts. So, when the voltage at IN is +5V, the voltage at the top of R2 should be +0.5V. The R1-R2 divider will have a resistance ratio of 9 to 1. To get the same voltage at OUT, the ratio of R3 to R4 should be 9 to 1. The ON/OFF CONTROL SIGNAL depends upon the voltage range that the gate of Q1 or Q2 can withstand. Assume that the transistor specification for the gate voltage (with respect to the source voltage) of Q1 and Q2 can be greater than +/- 10 Volts and that both Q1 and Q2 are fully ON when +5V (with respect to their emitters) is applied to their gates. Therefore, the minimum voltage of the ON/OFF CONTROL SIGNAL is the gate voltage at which the transistor (FET) saturates plus the maximum signal voltage:

-5V + -0.5V or -5.5V
and +5V + +0.5V or +5.5V

Because the input impedance of IC2 is very high, R5 is used as a current path for the signal through Q1 or Q2. The resistance of R5 should be more than 100 times greater than that of R2. R5 will then contribute less than a 1%, error to the signal voltage at the input of IC2. R5 is also used hold the input to IC2 at GND potential when Q1 and Q2 are turned off (ON/OFF CONTROL SIGNAL is low). In this case the known voltage is GND.

That's all I have to say.

Varying Current through a High Voltage Load with Low Voltage Control

Here's a circuit that uses the base-emitter junction of a high-voltage bi-polar power transistor to create a low voltage power supply. This supply is used by a low voltage device (bi-polar transistor or FET) to control the current through the high-voltage transistor. In the circuit below, the current is controlled by a Current Limiting Resistor and the current is either on or off. The value of R1 is selected to keep the High Voltage Transistor in saturation at all times. The High Voltage Transistor can be a power Darlington device but remember, the base-to-emitter voltage will increase by a diode drop.

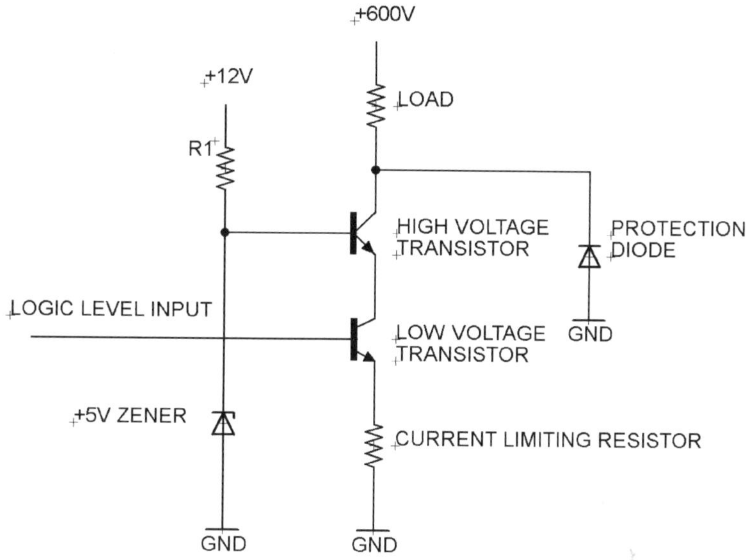

With a purely resistive load the protection diode is not necessary, but with any type of reactive load it is a good idea to have a current path around the high voltage device in order to protect it from excessive reverse voltage.

The High Voltage Transistor should probably be a TO-220 cased device or a TO-3 cased device, or something bigger. The Low Voltage Transistor can be a TO-99 or even a SOT-223 or SOT-23 surface mount device depending upon the load current.

O.K., o.k.! Maybe a power FET transistor is a better device for switching current through a high voltage. Just remember that the gate drive will have to produce enough current to charge a 100pF capacitance in just a few nano-seconds because you don't want the FET operating in it's linear region for very long. You'll also have to supply a large momentary current in the opposite direction in order to turn off the FET. Taking the 'gate' of the FET to a potential that is much less than the 'source' of the FET will help to make sure that the FET is definitely off (not conducting) but be sure that you don't exceed the negative gate-to-source specification of the device. Then too, you may want some type of current sense protection in order to turn off the FET in the event that the load current increases. The gate drive circuit becomes a little more complicated now.

There are trade-offs in everything. Good luck.

A List of Things that may Help

When troubleshooting, start with the power supply. Make sure you have input power and all the voltages.

Don't think you'll remember it (whatever 'it' is). Write it down in such a form that you'll understand it when you read it. This is especially true of software.

You don't know it until you can explain it. Teachers go through this all the time.

After you've had the fun designing, building and troubleshooting, don't forget to document, document, document! The best way is to write the production procedures.

Take the time to draw a flow diagram before writing a line of code. It doesn't have to be detailed, just use it to get your head organized.

Use an oscilloscope! You can't do it all with a multimeter. You miss so much when you don't use a scope.

Keep designing. Even when your job doesn't involve design, keep designing stuff. 'How do you get to a rewarding design? Practice, practice, practice.'

Learn several ways of solving a problem. A digital solution may be a lot cheaper and easier than an analog one, or vice versa. Mixing analog and digital often turns out to be the best solution.

If the specs allow, and the application is fairly simple, don't be afraid to use relay logic, off-the-shelf timers and counters or PLCs.

Follow your designs through to a working conclusion. To aid in this, learn at least how to lay out a double sided PC board and how to generate Gerber files. The turn-around time on prototype PC boards is in the neighborhood of three to four days in small quantities.

Even though you may need to buy some equipment, get involved in surface mount. With the right equipment, using 0805 sized components is no problem. 0603 isn't that bad either.

When a new chip doesn't seem to be acting the way you thought it should, you probably missed something in the data sheet. Never assume anything and don't be afraid to contact the manufacturer after you've exhausted every other option.

When designing high current circuits, be careful of ground loops (multiple paths to ground for the same current). These can cause oscillations that can waste a lot of good troubleshooting time, not to mention driving you completely bonkers.

If you can't meet a particular spec in your design, let everyone involved know about it. Go to the real decision makers and get the O.K. in writing!

A Capacitance Mirror Circuit

Here's another 'in case' circuit. Just in case you ever need to mirror the charge in a test capacitor and save it in a second capacitor or capacitors. Why? Well, you may only have access to the test capacitor through a path with varying resistance, such as a 'dirty' contact or a plasma.

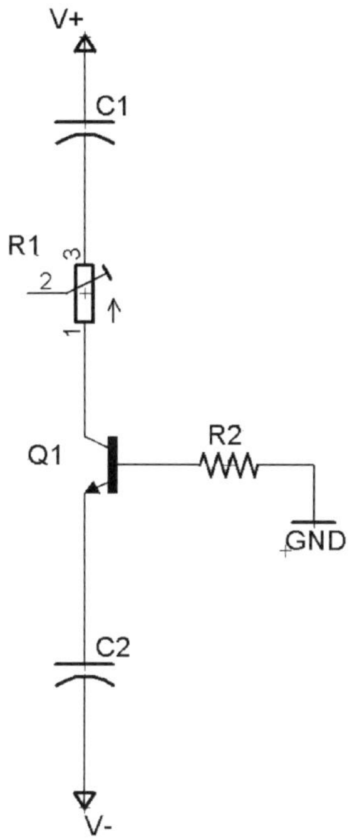

The basic concept is to use both a plus and a minus supply and a series pass transistor with an emitter voltage near ground potential. The idea is to have all the charge acquired by C1 to also be stored in C2. There's a big problem already with achieving this in the above circuit. A certain amount of drive current will pass through R2 and through the base emitter junction of Q1. This current will add charge to C2 but will not appear on C1, resulting in a mismatch between the charges on C1 and C2. In addition, any base current will cause a voltage drop across R2. This will result in a decreased voltage at the base of Q1, at the emitter of Q1 and across C2. So let's add an active component to try to keep the emitter of Q1 at ground potential. Use the V+ and V- supplies to power this component.

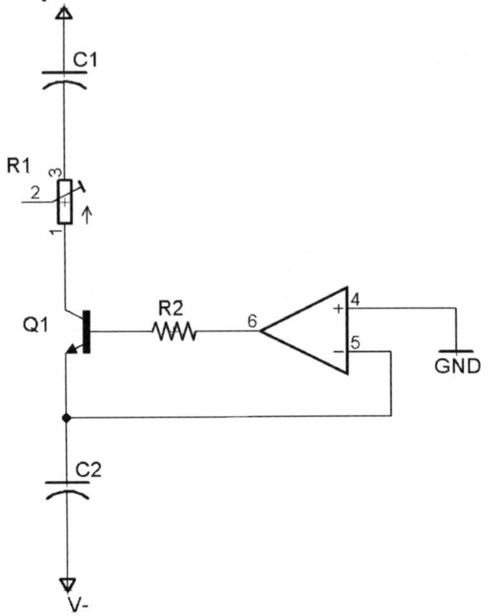

Well, this should keep the emitter of Q1 at GND potential. But we're still left with the problem of the base current through R2. So let's add a little more circuitry to compensate for this current.

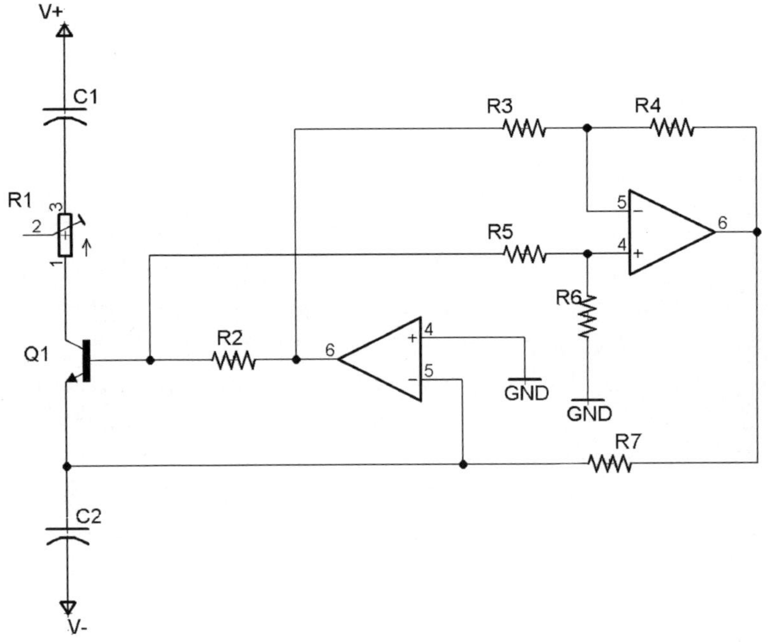

Aha! By using a difference circuit we can measure the voltage drop across R2 and produce an equivalent negative voltage across R7 for re-insertion of the lost current back into the series capacitive path. R2 and R7 will have the same resistance value and the output of the difference amp has the same voltage value as the drop across R2. The difference amp is configured to have a negative output.

Now, let's add some regulation to the current through the series circuit. A constant current circuit (Howland current pump) will provide a linear charge rate for the capacitor(s).

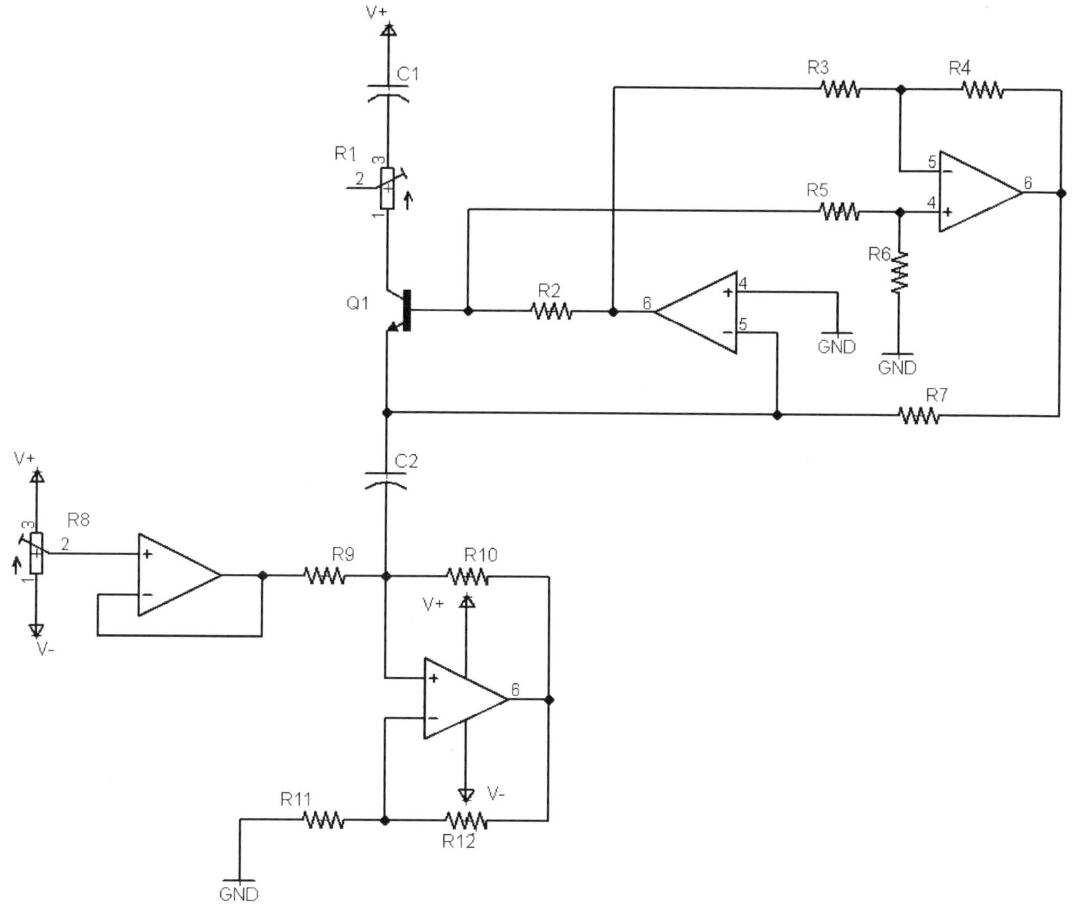

So there you have it. 'Have what?' you say. Well, there's still some work left. Like what to do with this mirrored charge in C2. Maybe you can add circuitry to discharge C2 through a constant current source and measure the time. Here are a couple of Physics equation that may come in handy:

$$CV = IT \quad \text{(Capacitance * Voltage = Current * Time)}$$ [1]

and

$$Q = CV \quad \text{(Coulombs of Charge = Capacitance * Voltage)}$$ [2]

Well, that's about it for this go-around. The circuits have been presented in no particular order, but that's usually the way designs come about. There are so many tools in electronics and somewhere along the way, you usually run into almost every one of them. You may like one more than another but give them all a chance. It's strange how things come around again. You'll probably wind up reaching into your bag of tricks for a solution to a new problem. Thanks for listening. Onward to the next great adventure.

END

[1,2] Fundamentals of Physics, Halliday, David; Resnick, Robert; Walker, Jearl July 1, 1999, Published by John Wiley & Sons ISBN 0471253820

Made in United States
Orlando, FL
04 November 2026

90204439R00125